EXCEL 2023

The Most Updated Bible to Master Microsoft Excel from Zero To Pro in Less than 5 Minutes A Day. Discover All the Formulas, Functions & Charts with Step-by-Step Tutorials, Tips & Tricks

Andrew J. Nash

GET YOUR BONUSES NOW!

The book on Excel will teach you how to use any function or chart but these Free Bonuses will make your life easier.

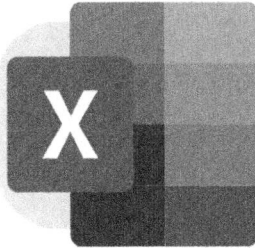

Excel Template

Follow this link to get Microsoft Excel templates

Excel Video Course

Follow this link to get Microsoft Excel Video Course

--- This BONUS is 100% FREE ---

SCAN THE QR CODE BELOW OR GO TO

Sommario

Introduction

In this eye-opening guide, we will go over some of the fundamental functions and steps of Microsoft Excel. It all comes down to making your job easier and more convenient. With the world changing, businesses are looking for employees who are not only familiar with the internet but also have a solid understanding of Microsoft Excel. Many people struggle with Microsoft Excel 2023 because it is a difficult technology to grasp and apply.

This user guide will teach Microsoft Excel 2023 in a simple and practical way. Upon completing this guide, you will be able to acquire the fantastic ability of Excel in no time. It will be beneficial if you set aside time each day so as to master Excel skills in one week. This book will cover the most important aspects of MS Excel 2023, including how to use it, its benefits, and other important features.

Microsoft Excel has become the industry standard for spreadsheet programming. Today, no other spreadsheet programming can compete with its popularity. It provides clients with an endless number of handy features and has been continuously upgraded and expanded over the years to assist people.

MS Excel is a simple program, and knowing the fundamentals can help students and professionals advance their careers. Microsoft Excel can help you manipulate, monitor, and interpret results, allowing you to make better decisions and save time and money.

The idea behind the book is a relatively simple concept. To comprehend fully, dedicating ample time to preparation is imperative. This will enhance your proficiency in using Microsoft Excel and boost your self-assurance to start utilizing the application independently. This way, you can create far more sophisticated programs capable of executing many tasks.

Furthermore, this book will help you learn a new skill, namely a thorough understanding of Microsoft Excel and you will also benefit from extensive practice. So, you'll be studying and practicing various ideas, functions, and formulas that you'll have created for your project or a tiny app that you'll use later in your personal, professional, or school life.

Everyone who wants to learn more about Excel 2023 should read this book. You don't have to be a great programmer to benefit from it. If you use MS Excel in your professional or personal life as a primary or secondary application, this book will provide you with a wealth of helpful information. This manual is intended for those who want to make their work easier and faster. Still, it may also require more advanced functions, such as managing databases, analyzing data, or having a dynamic worksheet available. With today's technology, you can do and redo calculations easily with Excel by just using a formula.

One of the prerequisites for you to use this wonderful program professionally and without too much difficulty is practicing. Do not give up when you get a bad result; try again and again. This is the only way to learn manual skills and understand how Excel functions and how to enter data so that the program works exactly the way you want it. By the end of this book, you will be astounded by the new features in the most recent version of Excel and how to use them. Be sure you read this book carefully, so you don't overlook any steps.

Let's get started!!!

PART ONE: EXCEL FOR BEGINNERS
Chapter #1: Introduction to Microsoft Excel

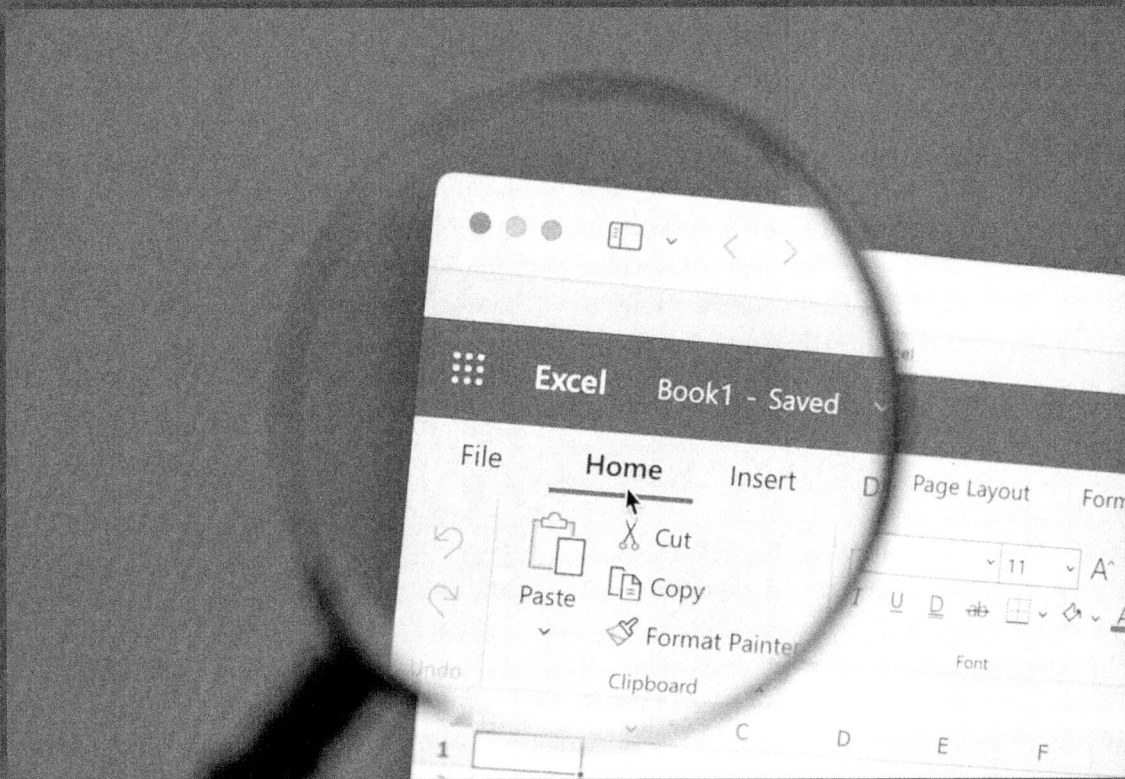

What is Microsoft Excel?

Excel is a spreadsheet program that organizes arithmetic operations using a grid of cells organized into numbered rows and letter-named columns. Many statistical, engineering and financial functions are included in the software. It can also display data in line graphs, histograms, charts, and limited three-dimensional visualization.

It allows users to see different perspectives on data by using pivot tables and the scenario manager (depending on various factors). Data is analyzed using pivot tables. It accomplishes this by simplifying data sets with pivot table fields. It also includes a programming component, Visual Basic for Applications, which enables the user to use a variety of numerical methods, such as those used to solve differential equations in mathematical physics.

There is also a variety of interactive components that allow for interfaces that conceal the spreadsheet entirely so that it appears as a decision support system (DSS) through an interface designed specifically for it, such as a stock analyzer or, in general, a tool that asks questions and provides answers.

You can use Excel to retrieve data from databases and measurement devices regularly, analyze the data, create a Word report or PowerPoint presentation, and send the presentation to a list of people. Excel's purpose is not only to be used as a database.

Several optional command-line tools provided by Microsoft can be used to control the launch of Excel. MS Excel and other spreadsheet software organize and manage data using rows and columns of cells. Excel users can arrange data in various ways, allowing them to examine various issues from multiple perspectives. Visual Basic for Excel is a programming language that will enable users to create complex numerical algorithms. Programmers can write, debug, and organize code modules directly in the Visual Basic Editor (VBE), which is included with Windows.

MS Excel's History

Microsoft Excel has been around since 1982 when it was first introduced as Multiplan, a popular CP/M (Control Program for Microcomputers) application that trailed Lotus 1-2-3 on MS-DOS platforms. In 1988, Microsoft released Excel v2.0 for Windows and began outselling Lotus 1-2-3 and developing QuatroPro. In 1993, Microsoft introduced Excel v5.0 for Windows, equipped with VBA (Visual Basic for Applications), commonly referred to as Macros. This created nearly limitless opportunities for enterprises to automate repetitive operations such as number crunching, process automation, and data presentation.

Microsoft Excel remains the most familiar, adaptable, and widely used business program, even with the introduction of Excel 2019 and Excel 365, owing to its ability to adapt to virtually any business process. There is nothing this compelling combination cannot handle when combined with other Microsoft Office software such as Word, Outlook, and PowerPoint.

In the beginning stages of accessible business computing for personal computers, Microsoft Excel played a crucial role in accounting and record-keeping for company operations. A spreadsheet with an automatic sum formula is one of the finest illustrations of utilizing MS Excel. With Microsoft Excel, entering a column of data, clicking into a cell at the end of the spreadsheet, and using the "auto sum" option to enable that column to add up all of the values entered above is extremely simple. This takes the place of manual ledger counts, a time-consuming aspect of the business before the invention of the modern spreadsheet.

Microsoft Excel has become a must-have for many forms of corporate computing, such as analyzing daily, weekly, or monthly numbers, tabulating payroll or taxes, and performing other equivalent business activities thanks to the auto sum feature and other enhancements.

Microsoft Excel has emerged as a prominent end-user technology that is useful in training and professional development due to a wide range of straightforward application scenarios. Because Microsoft Excel has been included in basic business diploma courses on computer-assisted drafting for many years, temporary employment firms may assess people's skills with Microsoft Word and Microsoft Excel before hiring them for a variety of clerical jobs.

Why Learn Microsoft Excel?

We all work with numbers in some way. We all have daily expenses that we cover with our monthly earnings. To spend wisely, one must first understand their income versus their spending. When it comes to recording, analyzing, and storing such quantitative data, Microsoft Excel comes in handy.

Excel's main function is to manage, organize, and perform financial analysis. Excel is used for anything involving calculations, business, accounting, and so on.

Excel can be used for a variety of tasks, including:

- Entering data
- Data Administration
- Accounting
- Financial Evaluation
- Graphing and charting
- Programming
- Task and Time Management
- Financial Simulation
- Management of Customer Relationships
- Excel's formula options can also be used to perform calculations.

Chapter #2: Microsoft Excel

How to download Excel

You can download Excel at this link: https://www.microsoft.com/en-us/microsoft-365/excel
Or, from your laptop, go to Microsoft.com and click on Excel.
Click on Buy now and select the option that best suits your needs.

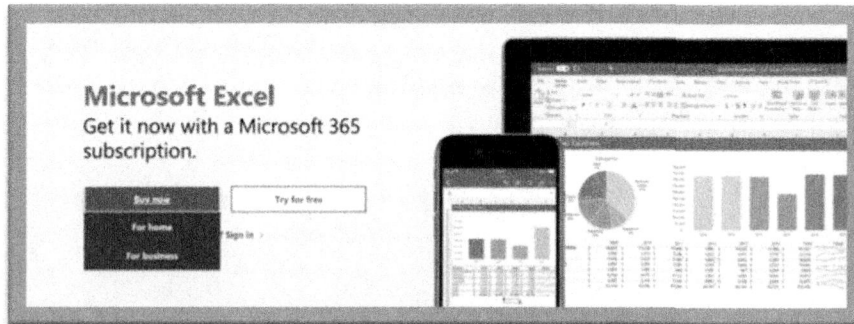

You can purchase multiple licenses and share them with your friends, or you can purchase a single license and pay it monthly or yearly to save 16%.

You can also start a FREE month-long trial if you don't want to spend any money during the initial phase and want to test the software more freely. If you download Office 365, your Excel version will work on desktop, mobile, and the web. By installing Office 365, you will gain access to the other software included in the Office Business Suite across multiple devices and operating systems.

Once you've decided on a plan, go ahead and purchase the license, download Excel, and install it. The installation procedure is standard, and we will not go over it in detail here. When the installation is finished, double-click the Excel icon to launch it.

How to Get Started

Microsoft Excel

Double-click on the Excel icon to launch the program.
This will open a new blank Excel File, where we will be working.

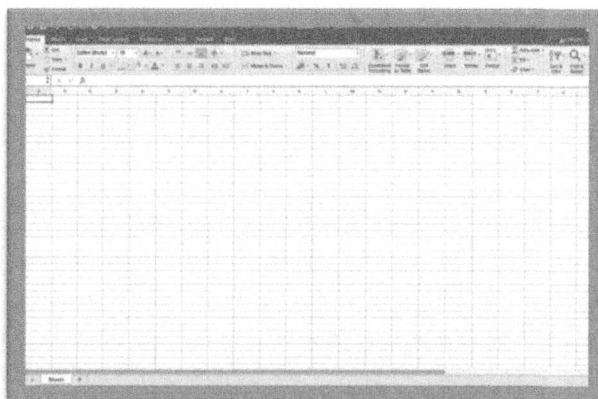

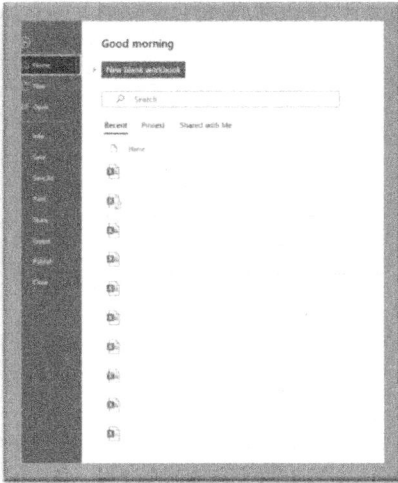

As you can see on the left, when clicking on File on the top left corner of the screen, you will be able to *save, name, open, and close* your workbooks anytime you need.

Excel does not start with a blank sheet when you launch it. You'll instead be taken to Excel's Main page, where you can load a current worksheet or choose a template. Excel displays worksheets that you've previously used, bookmarked or linked to you for easier access. If you have any worksheets that you frequently use, you can attach them to this Main page to make them easier to find. If you don't have a preexisting style, choose "New worksheet." Let take a short glance around it and examine some of its most frequent terms once you're on Excel's primary screen.

How to Get Excel

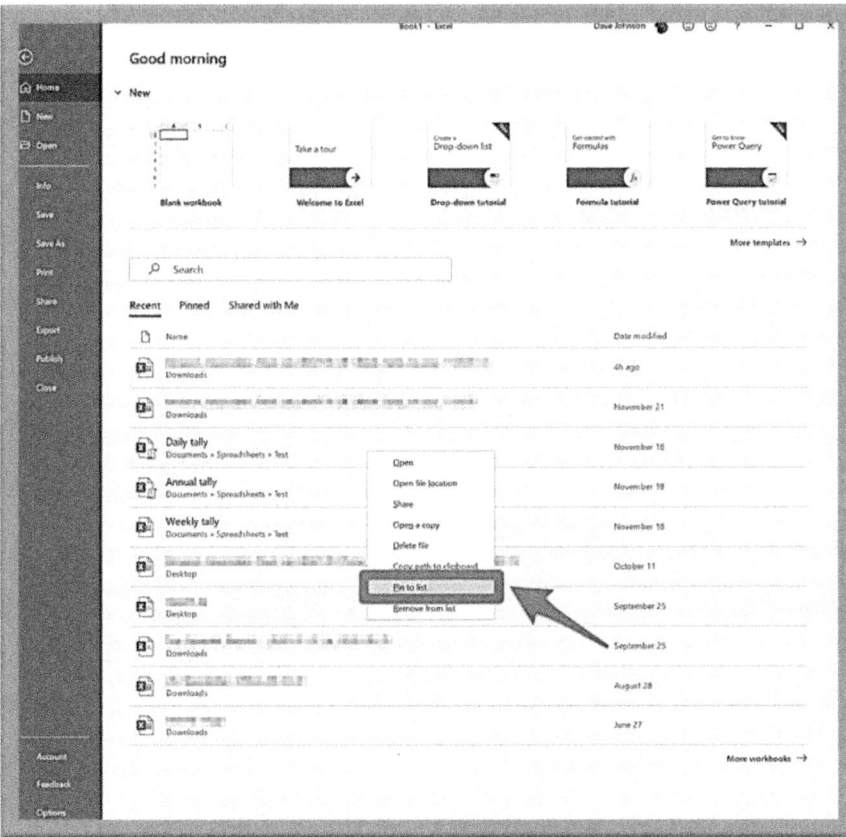

As a critical component of the enormously popular Microsoft Office operating system, Excel has become a tremendously desirable tool for organizations all over the world. While the Office 365 membership is excellent for businesses, it may not be suitable for you. Microsoft Excel and other key Office programs are available online for free; all users must have a Microsoft account. Visit Office.com and create a new profile or sign in to an existing one.

In the Website

Upon logging in to your Microsoft account, you will encounter a variety of applications at the front, including Excel, Word, and others. These are mostly 'Web Applications,' which imply software that is accessed via the internet rather than downloaded to your computer. When you work on a document it is securely saved to the OneDrive cloud platform. Indeed, Google deserves credit for the internet app version of this program as its low-cost Documents and Worksheets technology and integration with Google Drive made it difficult for Microsoft to continue charging customers. Excel's website is a much more condensed edition, so the comparison to Google Docs is more realistic than with the full PC edition. Nonetheless, both programs will meet the requirements of the average user.

On the Cellphone

MS Office mobile technologies are free to download on all current smartphones, possibly in response to Google. Microsoft Excel is available for both Android and Apple devices. Although Microsoft prudently reserves some premium services for its Office 365 membership, the phone applications remain competent and sufficient for working.

Through Subscription

In fact, if you need the full version of Excel, you must subscribe to Office 365. MS, like Adobe, prefers to charge a monthly fee to use their program rather than purchasing it outright. On the plus side, you will receive frequent upgrades, which will make it much safer than a single large annual release. The monthly membership fee starts at £6.99.

Microsoft Excel can be obtained in a variety of ways. It is available from a computer hardware store that also sells software. You can download it directly from Microsoft's site, but you have to buy the license key. You pay Microsoft to have the most recent Office suite installed on your computer using this method. By subscribing to this service, you will gain access to not only the Excel application, but also to other Microsoft applications such as Words, PowerPoint, Publisher, OneDrive, Access. You can also subscribe for a month or a year. The good news is that if you purchase this subscription, you will always have an up-to-date Excel application.

Let me teach you practically how to buy the Microsoft 365 suite.

First, visit the Microsoft 365 purchase page by visiting the link *https://www.microsoft.com/en/microsoft365* on your web browser. The link will take you to a page that lookslike the one below.

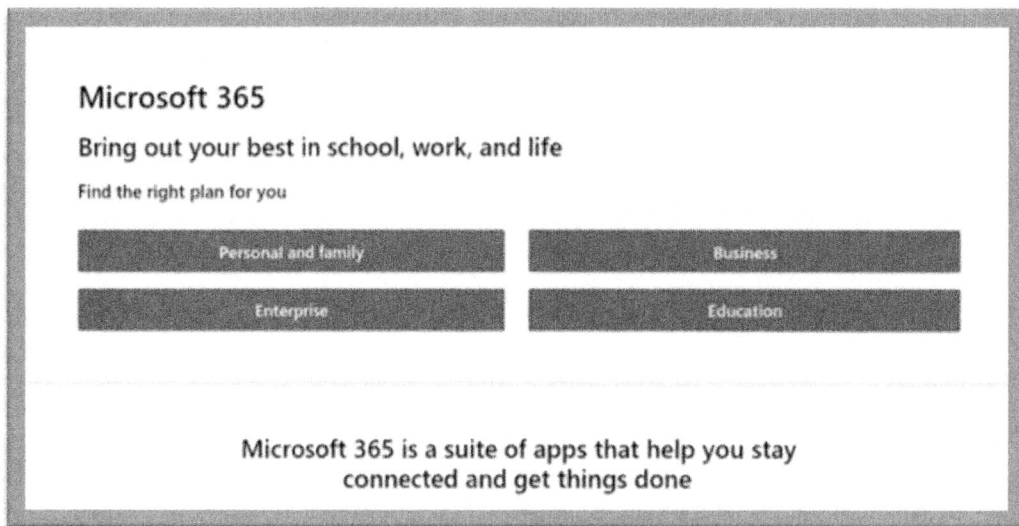

Microsoft 365 Purchase Page

As you can see in the photo above, there are categories of the suite you can go for. These are *Personal and family, Business, Enterprise*, and *Education*. For the purpose of this teaching, let me assume you are an individual, so, click *Personal and family*. As you take this step, a new page will open.

The new page will explain the features of the subscription you want to go for. The information includes some Microsoft apps you will have on your computer purchasing the package, OneDrive storage space, and how many people that can make use of the package. Be aware that there are other features that you will see on the page.

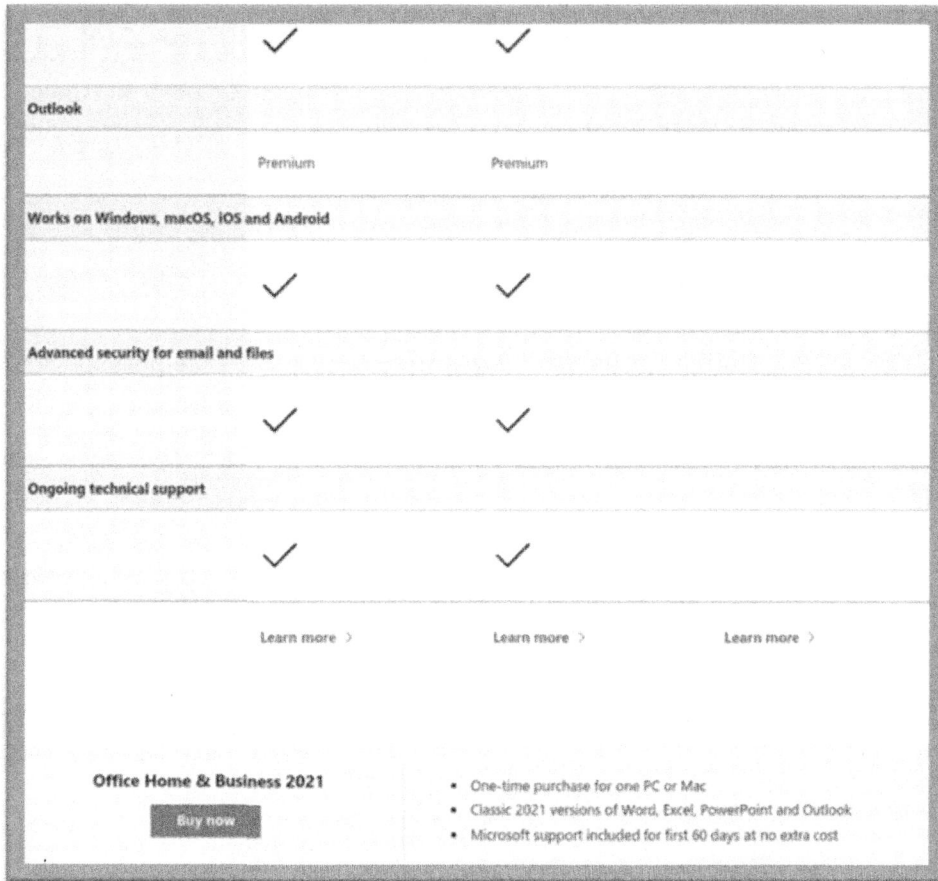

Progress in Buying Microsoft 365 Home Plan

The next step is to click the *Buy now* link. You will be taken to a checkout page. On the checkout page, click the *Checkout* link.

If you are not logged into your Microsoft account automatically, you will be required to insert your Microsoft email address, phone number, or your Skype ID. Insert any of the information and move to the next page. Insert your Microsoft account password and you will be logged into your Microsoft account.

Select the subscription duration you need and choose a payment method. The Microsoft 365 application software will be downloaded to your computer. Locate the downloaded file on your PC and double-click it, making sure to allow permission for it to make changes to your computer. Follow the prompt on your computer screen to complete the installation.

Once the installation of the Microsoft 365 is finished, you will see Excel as one of the component apps on your computer. If the Excel app is not pinned on your computer's taskbar automatically, just locate it by clicking the *Start* button on your PC and then search for Excel. After you've found it, select it to launch it. These are the steps to install the Microsoft 365 suite on your computer, of which Excel is one of the significant apps.

Chapter #3: Fundamentals

Customization Microsoft Excel Environment

Many People Prefer a Black Color Setting

If blue is your favorite color, you can change the change the theme color to blue. You won't need to use developer ribbon tabs if you are not a programmer.

There are lots of possibilities thanks to customizations. Here are someof the possible changes:

- Customizing the ribbon.
- Choosing a color scheme.
- Formula settings.

- Settings for proofing.
- Save the settings.
- Save Your Preferences

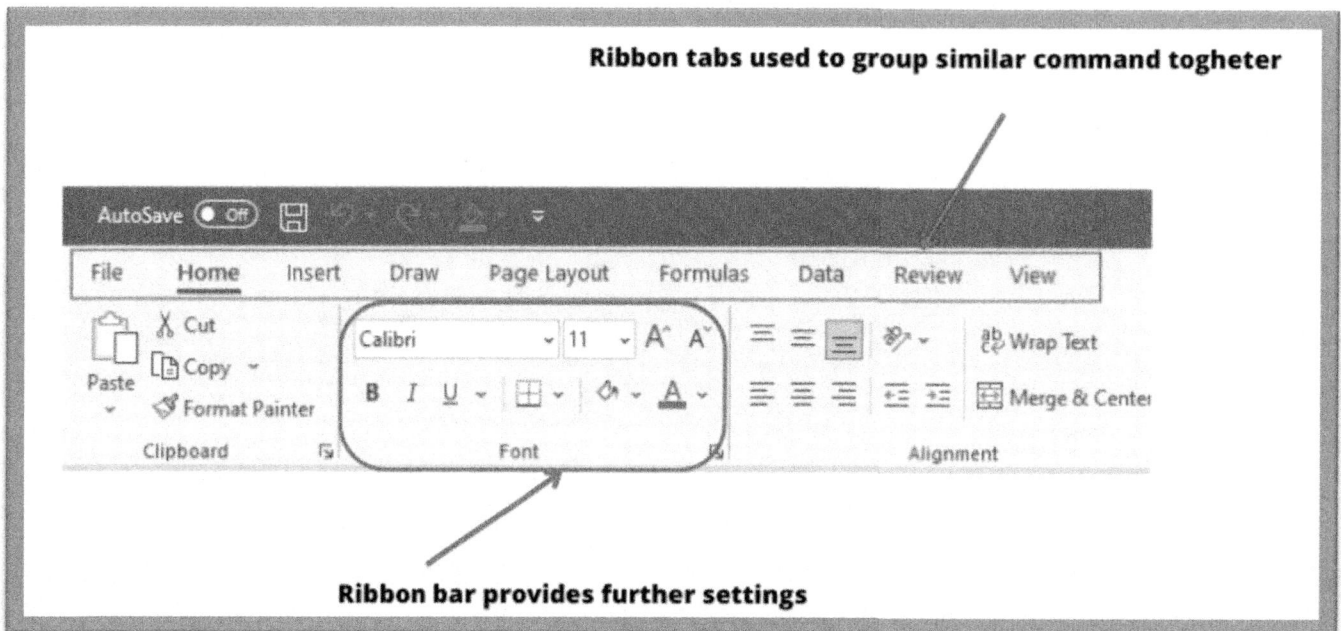

The Excel Tabs

When you launch Excel and select Blank Workbook or any existing template to begin building your data, one of the first things you'll notice as the workbook opens is the tabs. The Excel spreadsheet tabs are located at the top of the workbook interface. The photo below indicates the major tabs available by default in the Microsoft Excel workbook.

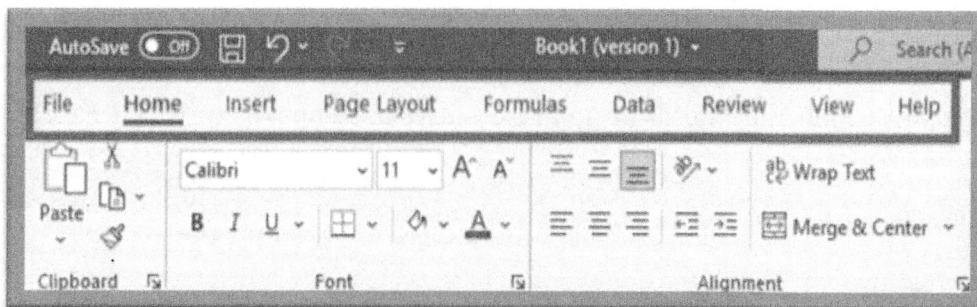

The Major Tabs in Rectangle

In the photo, the tabs are File, Home, Insert, Page Layout, Formulas, Data, Review, View, and Help.

Customization of the Ribbon

Let's start with the ribbon personalization. Let's say you don't want to see any of the toolbar tabs or you want to include some extra tabs, such as the developer tab. You can achieve this by utilizing the options window.

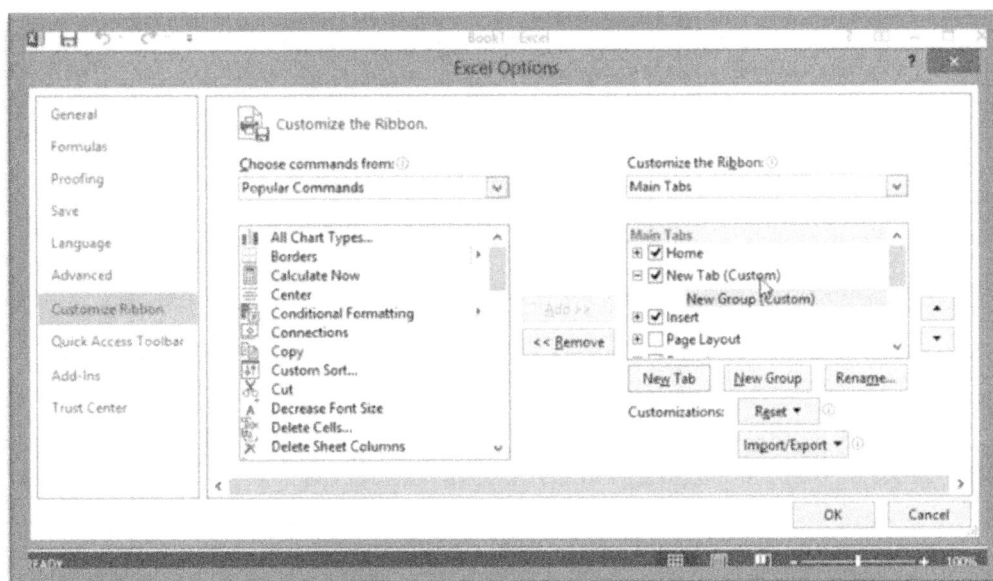

1. Select the start button on the ribbon.
2. Select options from the dropdown menu. A dialogue box called Excel options must be visible.
3. Select the customize ribbon option from the left-hand screen, as shown below.
4. Delete the checkpoints from the right-hand tabs that you don't want to see on the ribbon. This example does not include review, display tabs, or page layout.
5. When you're done, click the "Ok" icon.
   One may create their tab, call it whatever they want, and allocate it commands. Let's make a tab with the text (anything you want) in the ribbon.
1. Pick *customize the ribbon* from the context menu by right-clicking on the ribbon. A discussion window similar to the one seen above will emerge.

2. Click the new tab icon.
3. Go to the newly created tab and select it.
4. Choose the rename option.
5. Assign it the name ————
6. Under the ——— tab, choose a new group (custom), as seen in the image below.
7. Click the rename icon and assign it to my commands.
8. Let's move on to adding commands to the ribbon bar.
9. On the center panel, one will see a list of commands.
10. Press the add button after selecting the all-chart styles command.
11. Choose ok.

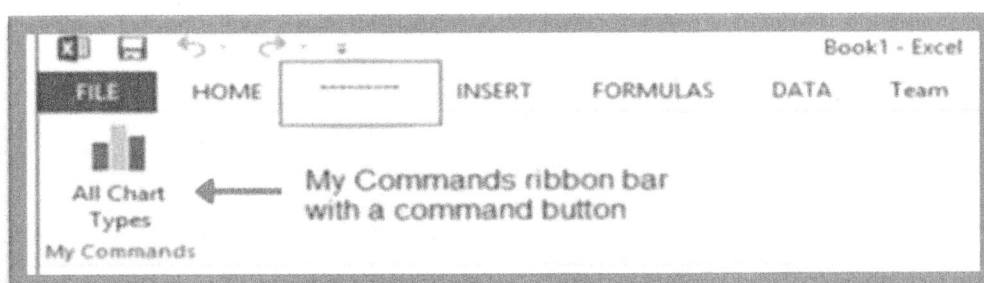

This is How the Ribbon Would Appear

How to Hide Excel Ribbon

To hide the Excel spreadsheet ribbon, first, click the *Ribbon Display Options* button

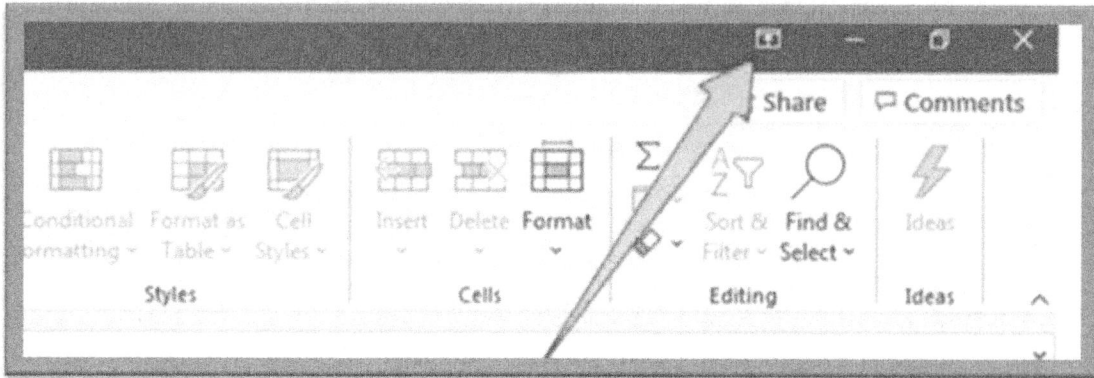

An arrow points to the Ribbon Display Options button

If your computer is running Windows 11 Operating System, the Ribbon Display Options icon's position looks slightly different. The icon is at the extreme top right of the Excel interface. The position is indicated in the photo below:

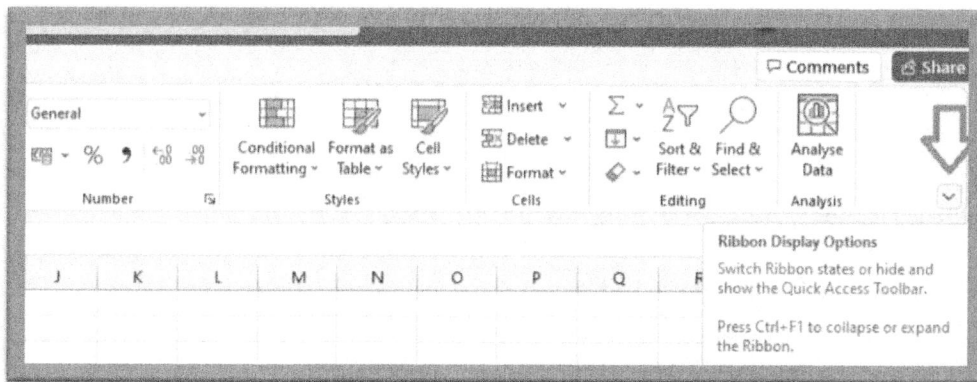

The position of the Ribbon Display Options Button in Windows 11

On clicking the Ribbon Display Options, you will see some commands displayed. This is shown in the photo below.

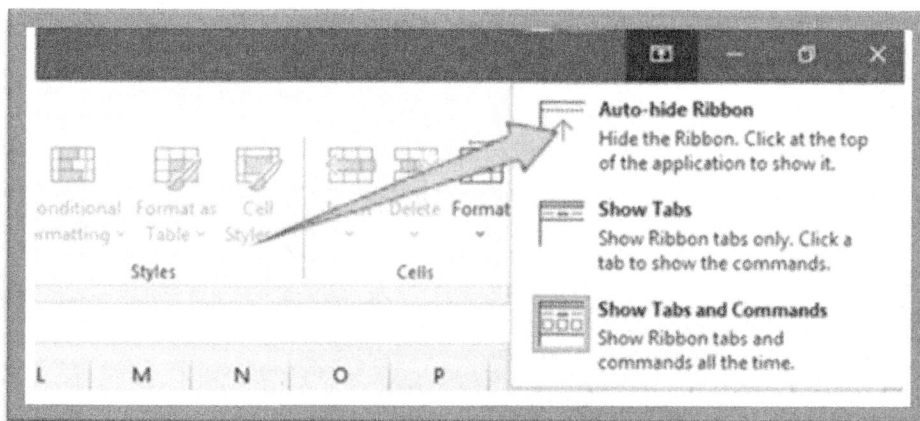

About Select a Ribbon Option

From the options, select *Auto-hide Ribbon*. When you select this option, all the major tabs and commands available on your spreadsheet app by default will not be there again. The photo below is how the new interface will look.

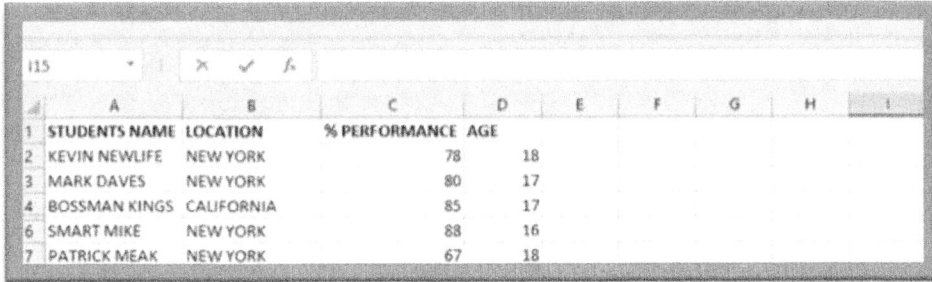

The new Interface of Microsoft Excel Software after Auto-Hide Ribbon ***Is Chosen***

The above screenshot shows that the major tabs and commands made available by default have disappeared.

Guide on How to Display Only Tabs on Excel Spreadsheet

Some Excel users choose to display only the major tabs on the ribbon section of the interface instead of having both tabs and commands the way they are made available by default.
They they prefer this because it gives them more space to work wiyh.
To display only tabs, click the *Ribbon Display Options* button at the top-right.

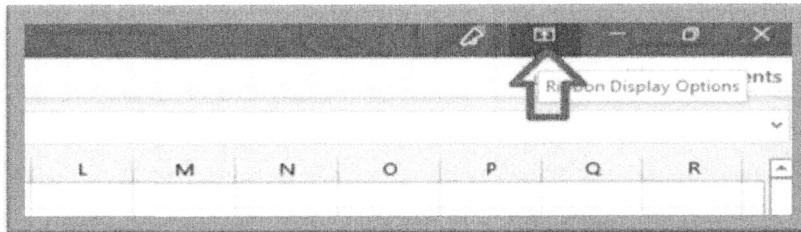

The Ribbon Display Options Button Pointed By the Arrow

On clicking the Ribbon Display Options button, you will see some options. Select *Show Tabs*. On selecting the *Show Tabs*, you will see that only the spreadsheet tabs appear, as shown below.

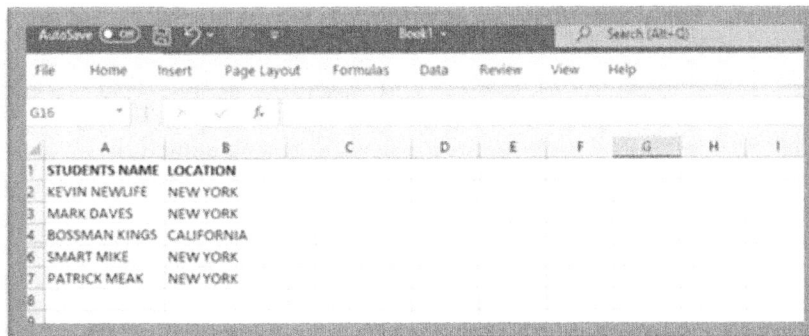

The Interface of the Excel Spreadsheet on Selecting Show Tabs

From the above photo, the tabs displayed are File, Home, Insert, Page Layout, Formulas, Data, Review, View, and Help, without including the commands under each tab. This will take us to the next subheading.

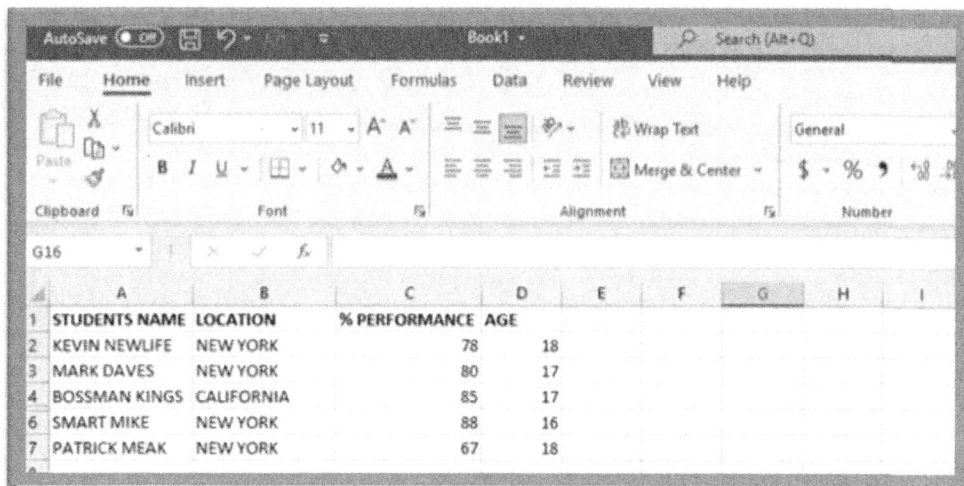

Both Tabs and Commands Are Displayed

How to Display Both Tabs and Commands on a Spreadsheet

To do that, click the *Ribbon Display Options* button. This action will display three options for you. Select *Show Tabs and Commands* option. In doing this, your spreadsheet will display both commands and tabs. You will see the interface appear below.

Using Rows and Columns for Navigation

Adding Columns and Rows

When you insert a formatted row, column, or cell into a worksheet, the Insert Options button appears. You will be presented with a number of formatting options for the inserted row or column. The options are summarized below.

- The format is the same as described above.
- The new row is formatted in the same way as the row above the inserted row.
- The format is the same as in the example below.
- The new row has the same formatting as below the inserted row.
- The same as on the left.

- The new column has the same formatting as the column to the left of the inserted column.
- The same structure as the right
- The new column has the same formatting as the column to the right of the inserted column.
- Formatting; basic formatting
- New rows and columns are formatted in the default manner.

To add a column:
1. Select *Insert* from the right-click menu of a column header.

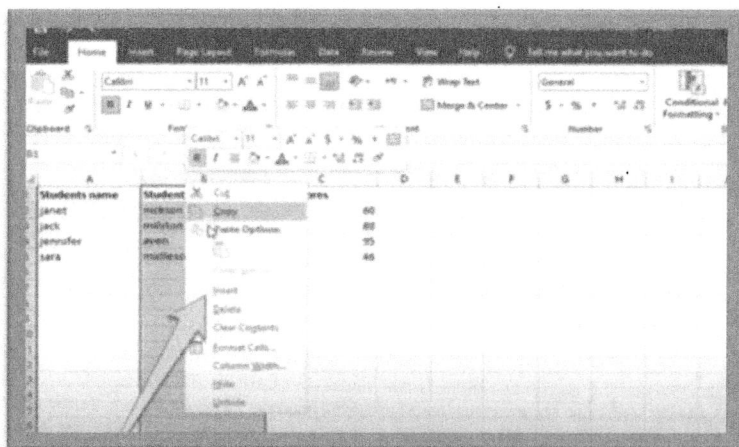

To insert multiple columns:

1. Insert *the same number of columns* as the number of column headers you selected.
2. Choose *Insert* from the context menu by right-clicking any column header.

To insert a row:
1. Select *Insert* from the right-click menu of a row header.

To insert multiple rows:
1. Insert *a row header* for each number of rows you wish to insert.
2. Select any row header, then choose *Insert* from the right-click menu.

Converting the Rows and Columns

When working with Excel, there are times when we need to convert columns into rows and rows into columns. This is referred to as transpose. Copying and pasting data from a row into a column would be time-consuming and inefficient.

So, in this case, one can select the transpose option, and rows will be converted into columns and columns into rows in no time.

Begin by selecting the columns and rows that you want to rearrange. From the context menu of the right-clicked item, select "Clone." Then, in your worksheet, select the cell where you want to begin your first row or column. After right-clicking on the cell, select "Paste Custom" from the context menu. A component will appear, and you'll see a transposition option near the end. Select that checkbox and press OK. This will convert columns into rows and vice versa.

You Can Also Split Data from One Cell to Another

What if users want to split data from a single cell into two distinct cells? Assume you want to find a person's company name based on their email address. Alternatively, for email campaign designs, you may want to divide a person's identity into the same first and last surname. Excel makes both possible. Begin by selecting the columns you want to separate. Afterward, go to the Information menu and select "Content to Column." A section with additional information will appear.

To begin, select "Delimited" or "Set Size."

The desire to segment the columns using symbols such as parentheses, periods, or spaces is "delimited."

"Set Size" indicates that you want to position the divide across all sections precisely.

In the scenario, we'll select "Delimited" to divide the full address into first and last names.

Following that, you must choose the Delimiters. It could be a space, a bar, a semicolon, a pause, or a gap.

Once you're satisfied with the display, click "Continue." This screen will also enable you to pick *Additional Templates* if you need it. When finished, click "Next."

Adjusting Columns and Rows

Cells can be inserted into a worksheet in the same way as rows and columns. In a column, you can use the Insert dialog box to shift the cells around the inserted cells down (in a row) or to the right (in a column).

Select the cells you want to move and point to the selection's border to move the data. Upon changing to a four-pointed arrow, you can drag selected cells to the target location on the worksheet.

Excel displays a dialog box when it detects that the destination cells contain data, asking if you want to overwrite them. This can be done by overwriting the data or by canceling the move.

To change the height of the rows:

1. Resize rows by selecting their headers.
2. Choose a row header and click the bottom border.
3. When the pointer changes into a vertical arrow, drag the border to adjust the row's height.

Or:

1. Rows can be resized by selecting their headers.
2. Right-click the selected row header and select Row Height from the menu that appears.

3. In the Row Height dialog box, enter a new row height.
4. Press the OK button.
5. To modify the column width
6. To resize the columns, select the headers.
7. Place the cursor on the right border of a column header.
8. Drag the column border to the desired width when a double-headed horizontal arrow appears.

Or:

1. Column headers for the columns you would like to resize should be selected.
2. To change the column width, right-click any of the selected column headers.

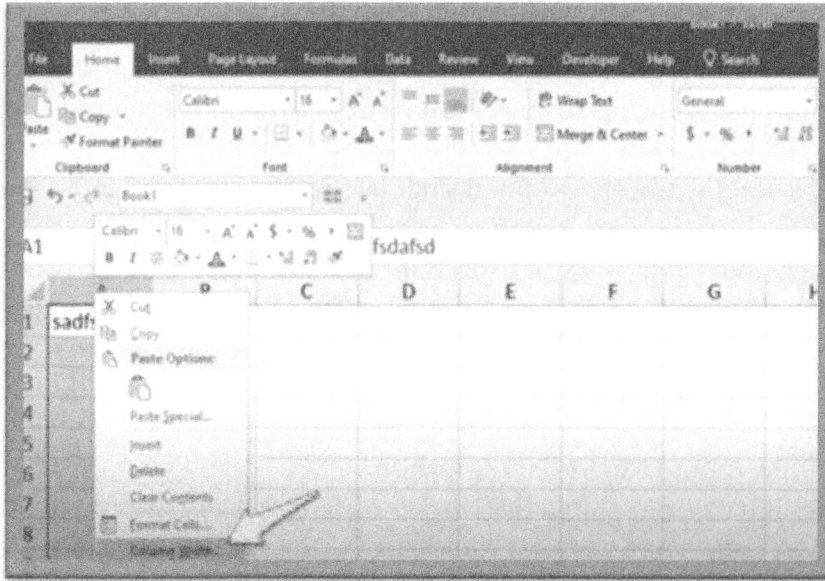

3. Fill out the Column Width dialog box with a new width for the columns you selected.
4. Click OK.

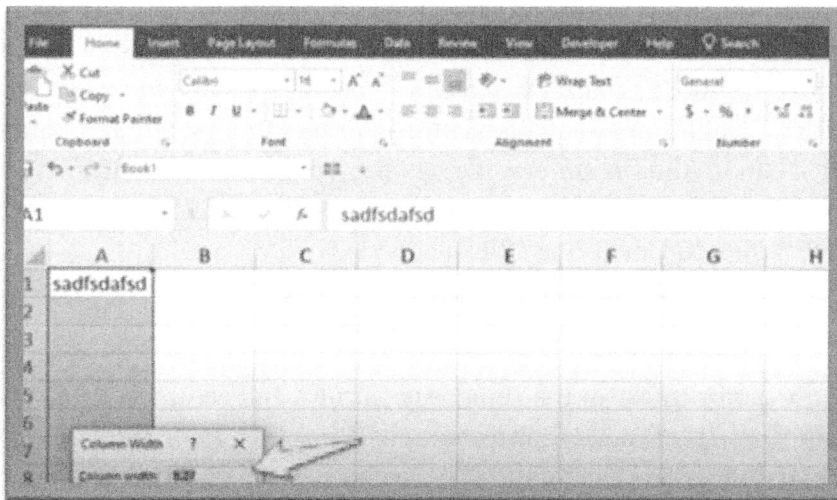

Working with Worksheets

1. Insert a new worksheet
2. Locate the new sheet key and click it.

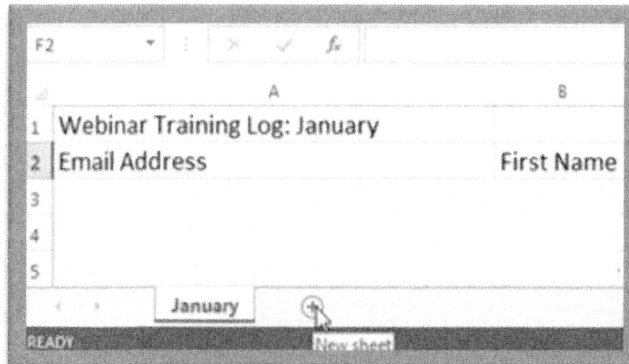

3. A new blank worksheet will appear on the screen.

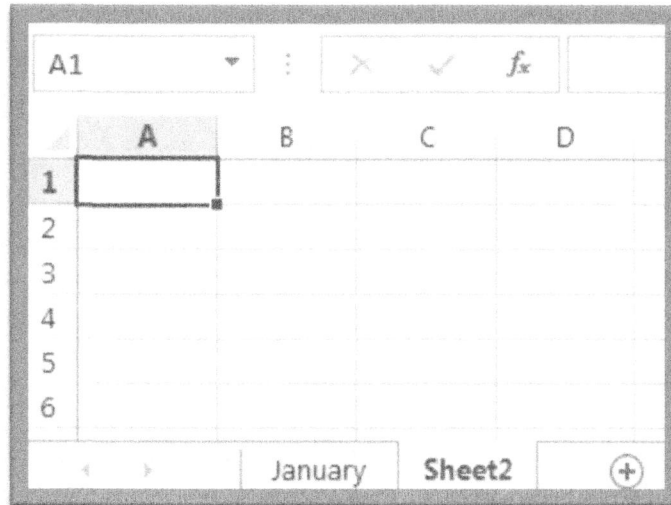

4. To modify the default workbook number, go to Backstage preview, tap Options, and choose the preferred number of worksheets in each new workbook.

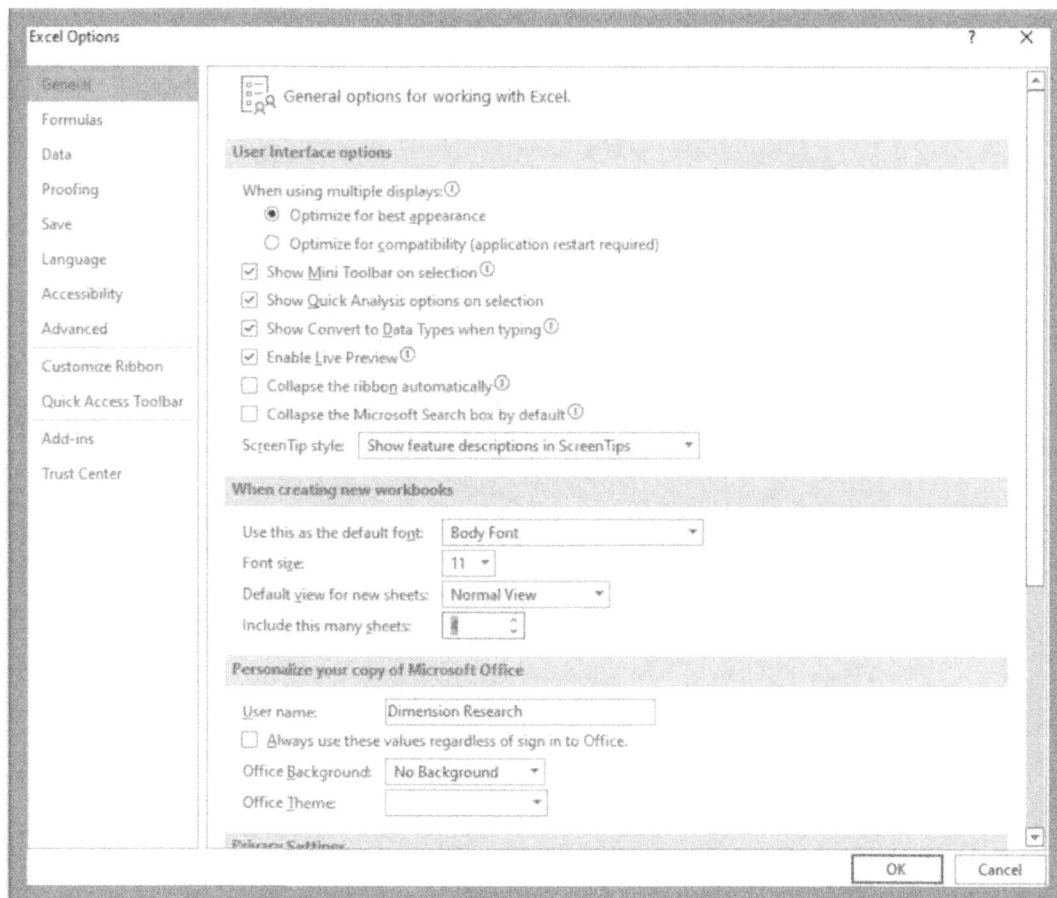

Rename a Worksheet

1. When you create a contemporary Excel workbook, one Sheet1 worksheet is supplied. A worksheet's name may be changed to represent its content better.
2. Choose Rename from the worksheet menu after right-clicking each worksheet you wish to rename.

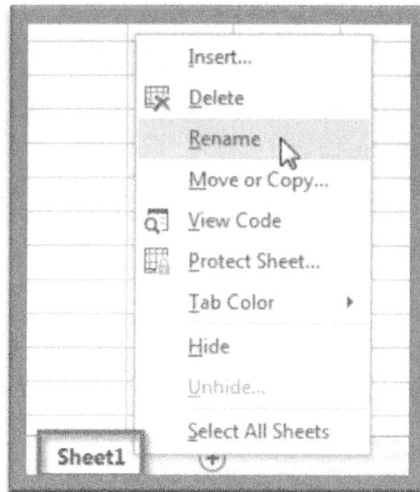

3. Fill in the name of the worksheet you want to use.

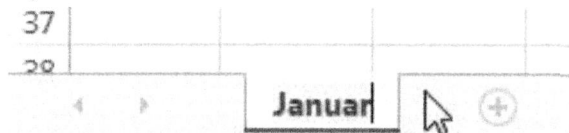

4. Press Enter or move your mouse outside of your worksheet using your keyboard. The worksheet's name will be changed.

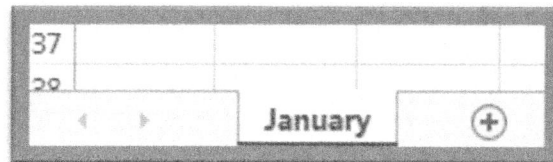

Delete Worksheets

1. Choose Delete from the worksheet menu after right-clicking a worksheet you wish to delete.

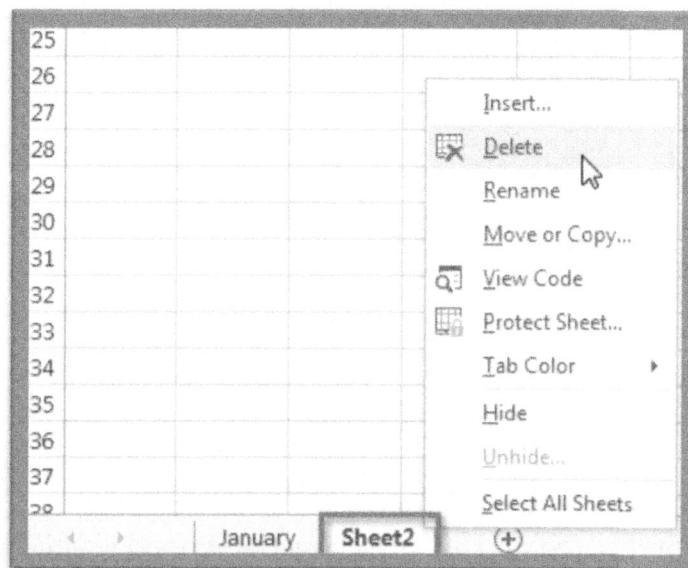

2. Your workbook's worksheet will be removed.

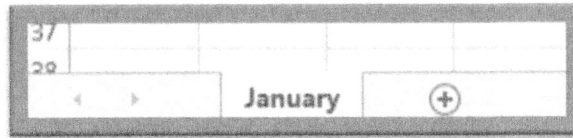

3.   Individual worksheets may be safeguarded from being altered or deleted by choosing Protect sheet from the worksheet menu when right-clicking the worksheet you wish to protect.

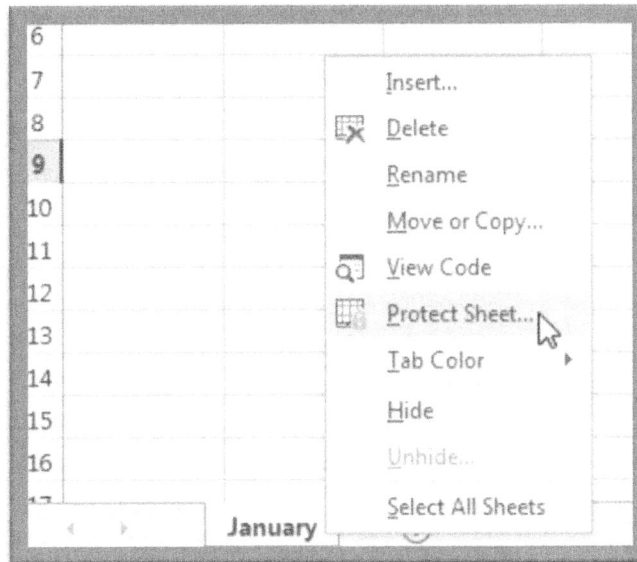

Copy Worksheet

1.   If you want to copy the contents of one worksheet to another, you can do so with Excel.

2.   After right-clicking any worksheet to copy, select Move or Transfer from the worksheet menu.

3.   The "Move / Copy" dialog box will appear. Indicate where your sheet should appear in the Before sheet: section. In this case, you'll move the worksheet to the right of the current one (move toward the end).

4.   Click OK after selecting Create a copy from the dropdown menu.

The worksheet will be duplicated. Because you cloned a January worksheet in your scenario, it will have the same name and version number as the original (2). The information from the original January worksheet will have been duplicated in the January (2) worksheet.

A worksheet can be copied to another workbook. From the book dropdown menu, you can select any available workbook.

Move the Worksheet

You may need to change worksheets to reorganize the workbook. Select the worksheet that you want to change. The pointer may transform into a small worksheet symbol. Maintain your pointer over the target area until you see a small black arrow.

Let go of the mouse button. The worksheet will be moved to a new location.

Colorize the Worksheet Tabs

1.   Change the color of a worksheet page to organize your worksheets and make your workbook more user-friendly. After right-clicking the corresponding worksheet tab, move the cursor over the Tab color. On the screen, a color menu will appear.

2. Choose a color that appeals to you. The most recent worksheet tab color sample will appear as you move your cursor over various options. We will use red as an example.
3. The color of the tabs on the worksheet will be changed.
4. When you select a worksheet, the color of the worksheet tab fades. Choose another worksheet to see how the color changes when the worksheet is not selected.

Changing Between Worksheets

By clicking the tab, you can navigate to a different worksheet. It can become tiresome when dealing with big workbooks, and you may need to go through all of the tabs to locate the one you are looking for. A better solution is to right-click the scroll arrows in the bottom left corner of your screen.

A dialogue box will appear with a list of all sheets in the workbook. Then double-click the sheet you want to jump to.

Worksheets for Grouping and Ungrouping

You can work on each worksheet individually or on several worksheets at the same time. Combining worksheets may result in the creation of a collection of worksheets. Any changes you make to one worksheet in a category affect all worksheets in that category.

To group the worksheets:

1. Imagine a scenario in which employees must be trained every three months. As a result, you'll make a worksheet category only for them. When you add an employee's name to a worksheet, it appears on all worksheets in the group.
2. To add a worksheet to a worksheet category, choose the first worksheet.

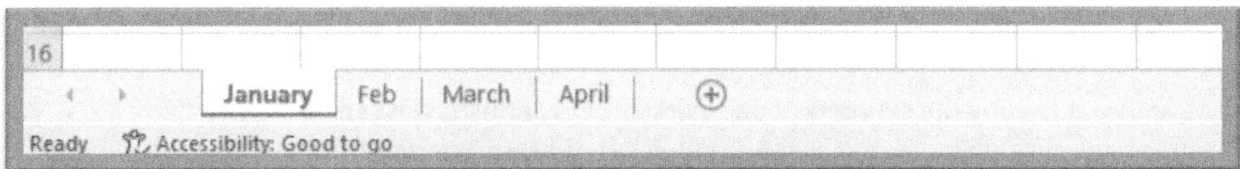

3. Hold down the Ctrl key on your keyboard.
4. Select the next worksheet for the group from the dropdown menu. Continue selecting worksheets until you've picked all the worksheets you wish to organize.

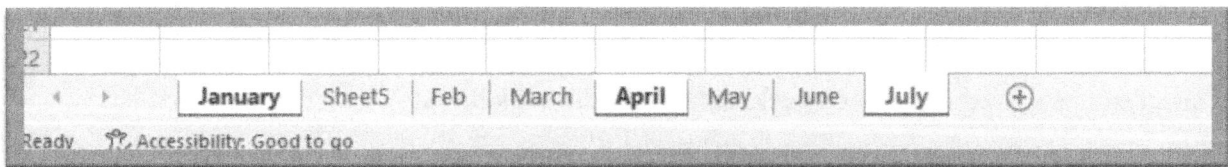

5. Release the Ctrl key. Your worksheets will now have been organized into groups.

You may explore a worksheet inside a category after it has been sorted. All modifications to one worksheet would be reflected in the rest of the group's worksheets. You'll have to ungroup all the worksheets if you want a worksheet that isn't part of the community.

To ungroup all worksheets:

1. You can obtain Sheets from the worksheet menu by right-clicking on any worksheet and selecting "Ungroup" from the menu that appears to separate all previously grouped worksheets.

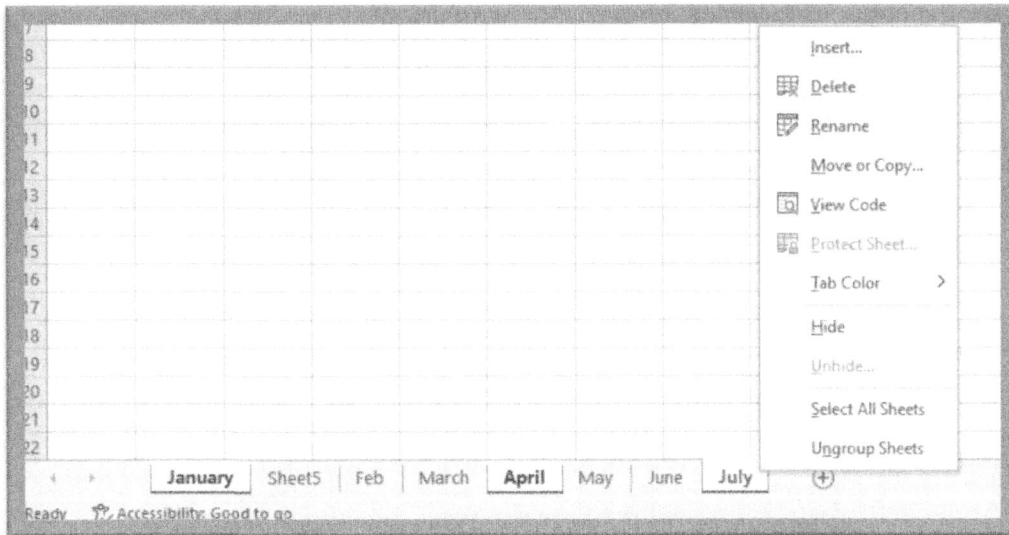

2. The worksheets will have been separated into groups. Instead, click on one of the worksheets that aren't part of the group to ungroup them.

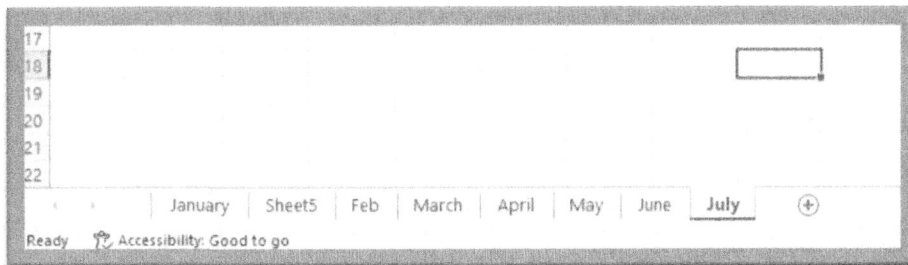

3.   Worksheets may be grouped and ungrouped. Group your January and March worksheets if you're following this particular scenario. After adding new material to a January worksheet, compare it to a March worksheet.

To hide a worksheet:

1.   To hide a worksheet, right-click its worksheet tab and select Hide.

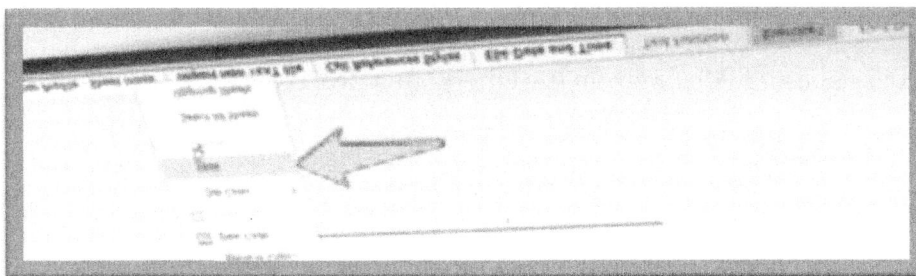

To unhide a worksheet:

1.   To unhide a hidden sheet tab, right-click it, then select *Unhide*.
2.   Choose the worksheet you want to display from the Unhide dialog.
3.   Click *OK*.

Modify Worksheets

A worksheet row is identified by a number, while a column is identified by a letter or a series of letters. Each row has a header on the left edge, and each column has a header at the top.

You can change the width or height of a worksheet's column or row by dragging the right edge of the column header or the bottom edge of the row header. By increasing the width or height of a column or row, you create more space between the content of adjacent cells, making the data easier to read and process.

Adjusting column width and row height and inserting a row or column between cells containing data can make working with workbook content easier.

When the content of a worksheet has space between its edge and cells containing data, or perhaps space between its label and the data to which it refers, the workbook will appear less crowded.

Page Layout view

You may wish to open the workbook in the Page Layout view to see your changes to the page layout.
The command to view Page Layout is located in the bottom right corner of your worksheet and can be accessed by clicking on it.

Page Orientation

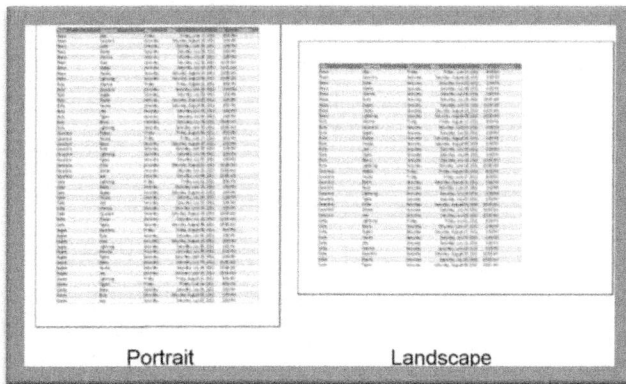

Portrait   Landscape

Landscape and portrait are the two-page orientations available in Excel. The landscape is horizontally oriented, while the portrait is vertically oriented. The portrait is the greatest option when working with many rows, whereas landscape is the best option for a bunch of columns. Portrait orientation works best when a spreadsheet has more rows than columns.

To change page orientation:
1. On your Ribbon, choose the Page Layout tab.
2. Choose between Landscape and Portrait from the Orientation dropdown box.

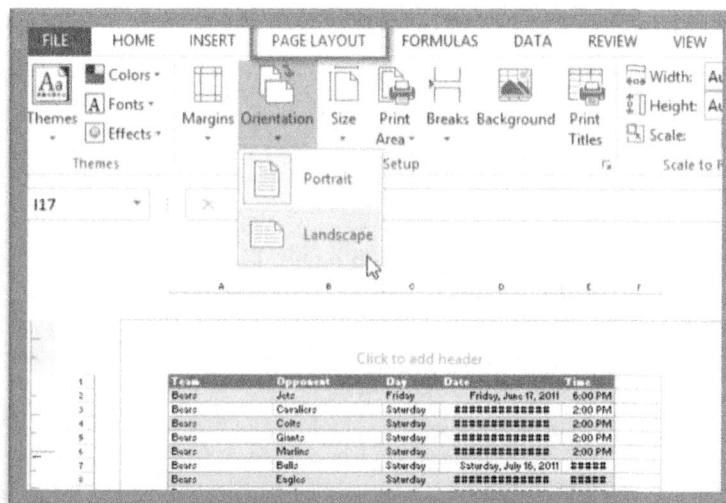

3. The workbook's page orientation will be altered.

To format page margins:

1. The margin is the space between the content and the edge of the page. The default workbook setting is a one-inch gap between the content and the page's edges. If your data doesn't fit on the page, you may need to adjust the margins. Excel has a number of predefined margin widths.

2. From the Page Layout menu on the Ribbon, select Margins to command.

3. From the dropdown menu, choose the appropriate margin size. If you want to include more on the page, you may choose Narrow.

4. In this case, the margins would be resized to fit your new selection.

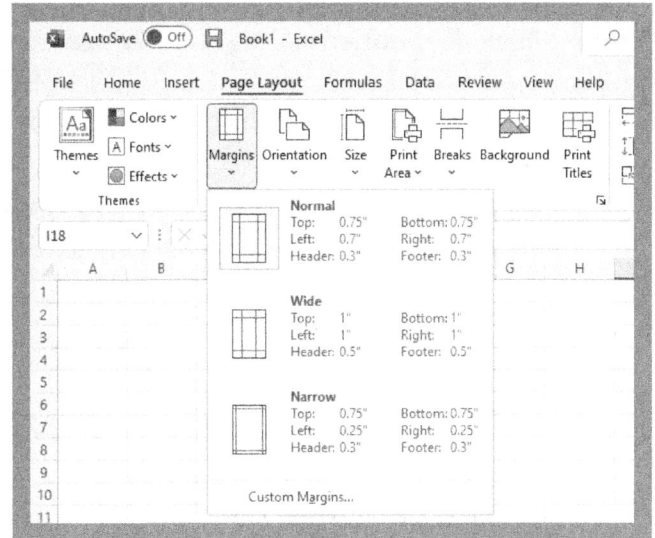

To use custom margins:

1. Excel allows you to change the margin size within the Page Setup dialogue box.

2. Click Margins on the Page Layout tab. The dropdown option will allow you to choose Custom Margins.

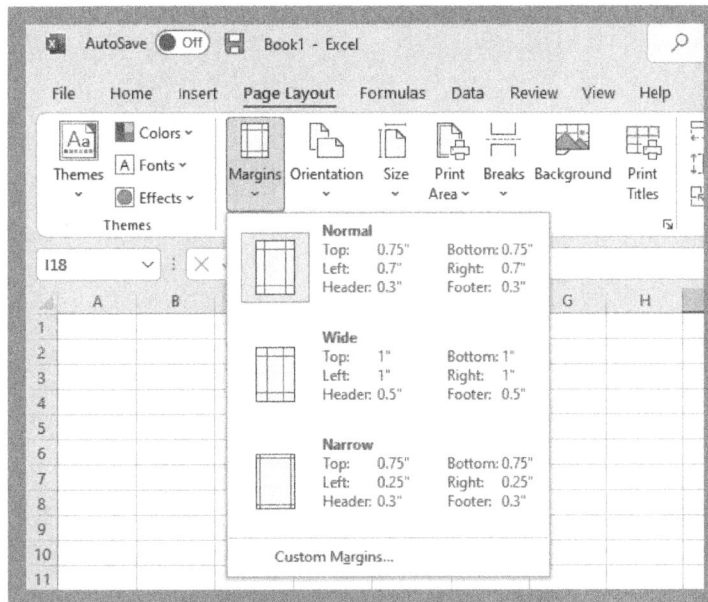

3. The Page Setup dialogue box will appear.

4. Then click the OK button to display your changes.

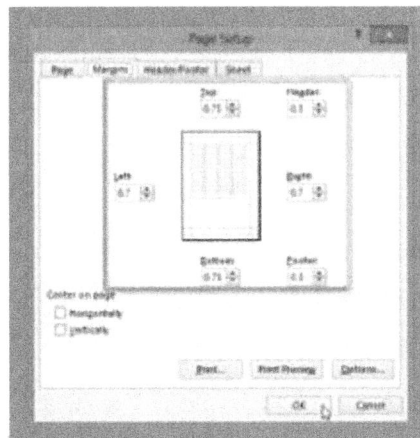

5. The notebook's margins will be adjusted.

To include Print Titles:

1.   Include all title headers on each page to ensure a complete printout of your worksheet. It would be difficult to understand the print if the title headers only appeared on the first page of a printed workbook. Print Titles commands allow you to choose which rows and columns to display on each page.

2.   Select Print Titles from the Page Layout menu on the Ribbon.

3.   Page Setup will now be shown for you to make changes to your page. Rows and columns may be repeated on each page. A row will be repeated in our case.

4.   Rows to repeat at the top may be found by selecting the Collapse Dialogs button.

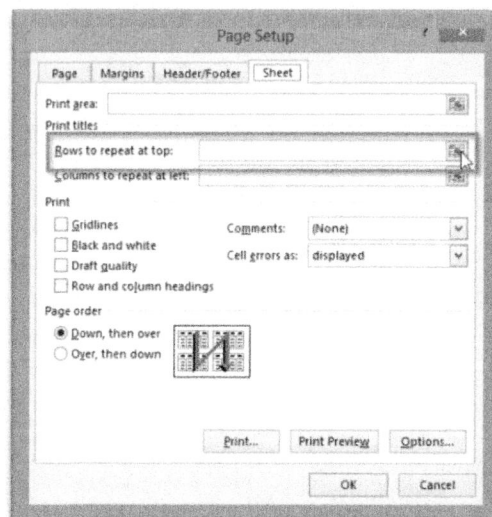

5.   A little selection arrow appears in place of the mouse pointer, and the Page Setup dialogue box is collapsed. Choose the rows you wish to repeat to have the same row on every printed page. For the sake of illustration, let's take row 1.

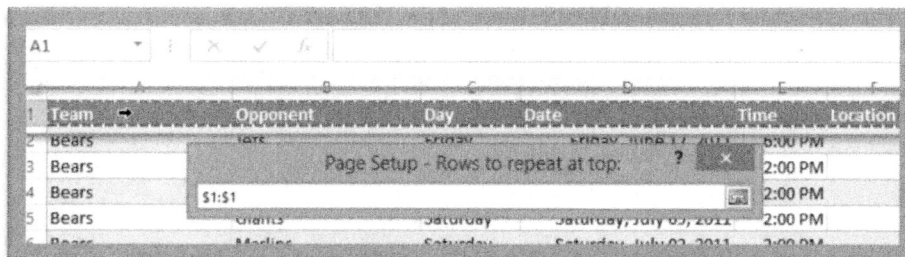

6.   In the Rows for repeating at the top field, Row 1 will be inserted. It's time to reactivate the "Collapse Dialog."

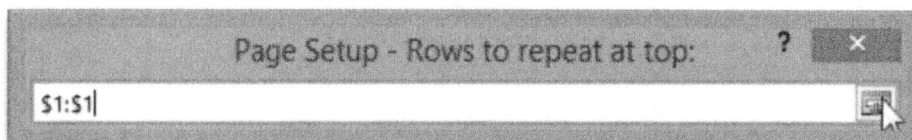

7. Expand the Page Setup dialogue box. Click the OK button to confirm your action. Every page will have Row 1 over top of it.

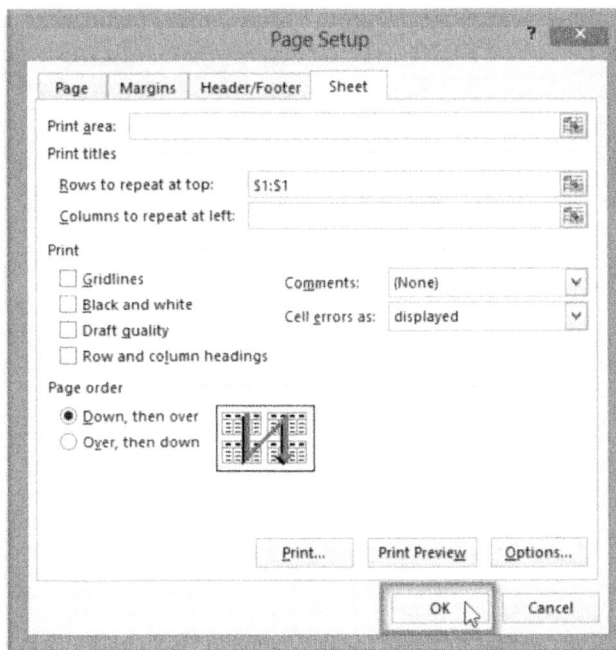

To insert a page break:

1.   You can print different document sections on different pages by inserting a page break into your workbook.

2.   There are both vertical and horizontal page breaks available. Horizontal and vertical page breaks separate columns and rows. In this example, we'll use a horizontal page split.

3.   This command for the Page Break view can be found and selected here. The worksheet will be displayed in Page Break mode.

4.   Click on the row beneath wherever you want the page break to appear and then choose the appropriate check box. A page break may be added by selecting row after row 28 (in this case, 29).

	A	B	C	D	E
19	Bulls	Lightning	Saturday	Saturday, June 18, 2011	10:00 AM
20	Cavaliers	Eagles	Friday	Friday, August 05, 2011	6:00 PM
21	Cavaliers	Hawks	Friday	Friday, June 17, 2011	6:00 PM
22	Cavaliers	Bears	Saturday	Saturday, August 13, 2011	2:00 PM
23	Cavaliers	Bulls	Saturday	Saturday, June 25, 2011	2:00 PM
24	Cavaliers	Lightning	Saturday	Saturday, July 16, 2011	2:00 PM
25	Cavaliers	Tigers	Saturday	Saturday, July 02, 2011	2:00 PM
26	Cavaliers	Colts	Saturday	Saturday, August 20, 2011	10:00 AM
27	Cavaliers	Giants	Saturday	Saturday, July 23, 2011	10:00 AM
28	Cavaliers	Jets	Saturday	Saturday, July 09, 2011	10:00 AM
29	Colts	Lightning	Friday	Friday, July 01, 2011	6:00 PM
30	Colts	Bears	Saturday	Saturday, June 25, 2011	2:00 PM
31	Colts	Eagles	Saturday	Saturday, August 13, 2011	2:00 PM

5.   Insert a page break by clicking your Page Layout tab and within that Ribbon selecting your Breaks command.

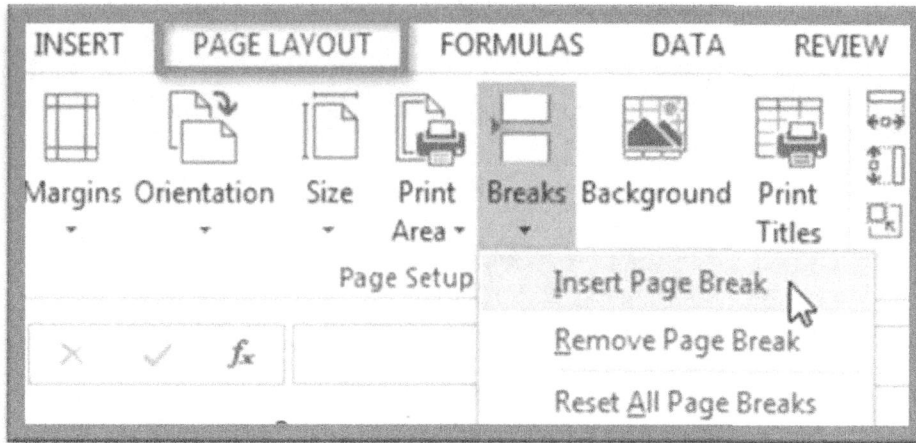

6. A dark blue bar denoting a page break will be placed.

	A	B	C	D	E	F
19	Bulls	Lightning	Saturday	Saturday, June 18, 2011	10:00 AM	
20	Cavaliers	Eagles	Friday	Friday, August 05, 2011	6:00 PM	
21	Cavaliers	Hawks	Friday	Friday, June 17, 2011	6:00 PM	
22	Cavaliers	Bears	Saturday	Saturday, August 13, 2011	2:00 PM	
23	Cavaliers	Bulls	Saturday	Saturday, June 25, 2011	2:00 PM	
24	Cavaliers	Lightning	Saturday	Saturday, July 16, 2011	2:00 PM	
25	Cavaliers	Tigers	Saturday	Saturday, July 02, 2011	2:00 PM	
26	Cavaliers	Colts	Saturday	Saturday, August 20, 2011	10:00 AM	
27	Cavaliers	Giants	Saturday	Saturday, July 23, 2011	10:00 AM	
28	Cavaliers	Jets	Saturday	Saturday, July 09, 2011	10:00 AM	
29	Colts	Lightning	Friday	Friday, July 01, 2011	6:00 PM	
30	Colts	Bears	Saturday	Saturday, June 25, 2011	2:00 PM	

A solid grey line indicates an added page break, whereas a dashed grey line signifies an automated page break while looking at your workbook within Normal mode.

To insert headers and footers:

1. Headers & footers enhance the readability of your worksheet and give it a more polished appearance. The header and footer are two sections of the worksheet that appear in the upper and lower margins, respectively. Page numbers, dates, and workbook names are common identifiers seen in headers & footers.

2. This command to view Page Layout may be found at the Excel window's bottom. You'll see the worksheet within Page Layout mode.

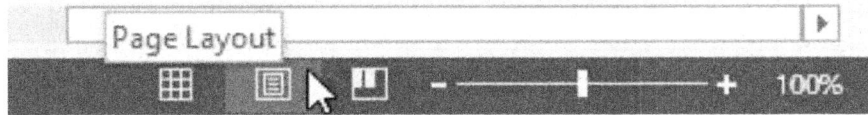

3. If you wish to change a footer or header, choose it. Changes will be made to the footer of this page.

Your Ribbon will display a new tab titled Header & Footer Tools. Commands that contain dates, page numbers, and workbook names may be found here. We add Page numbers in the example below.

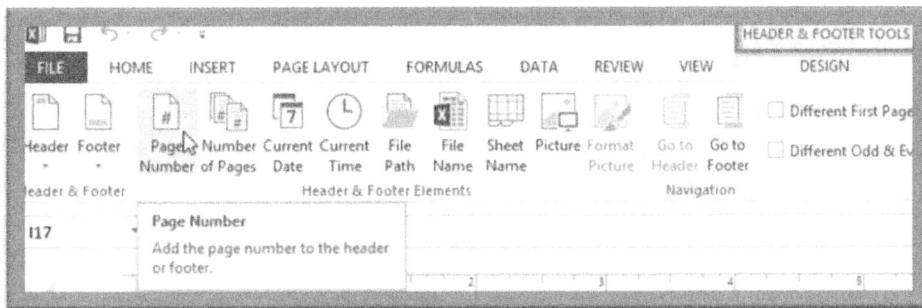

Create and Manage Workbooks

To Create a New Workbook
Suppose Excel isn't running:
1. Open Excel (without opening a particular workbook).

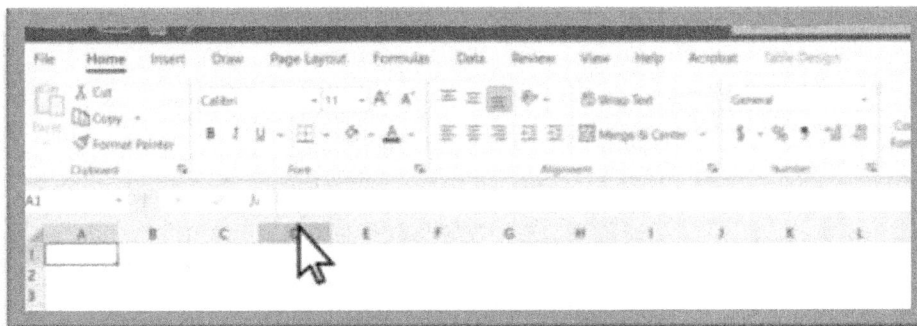

2. Choose the Blank Workbook option on the Start screen.
3. Show the new page from the backstage view when Excel is running.
4. Select the option to create a new workbook.

Creating a Workbook from an Existing Template
1. In the backstage view, display the new page.
2. Enter a search term in the "Search for online templates" box and press Enter.
3. You can select a template and then click "Create."

To Make a Backup Copy of a Workbook

1. You have two options for viewing the Save As page from the backstage view.
2. You can return to the original Save As dialog by pressing F12.
3. Navigate to the location where you want to save the workbook.
4. In the File name box in the upper-right corner of the Save As page or dialog, type in the new name for the workbook.
5. Select a different file type from the Save As type list to change the file format option.
6. Save the document.

To Open a Previously Saved Workbook

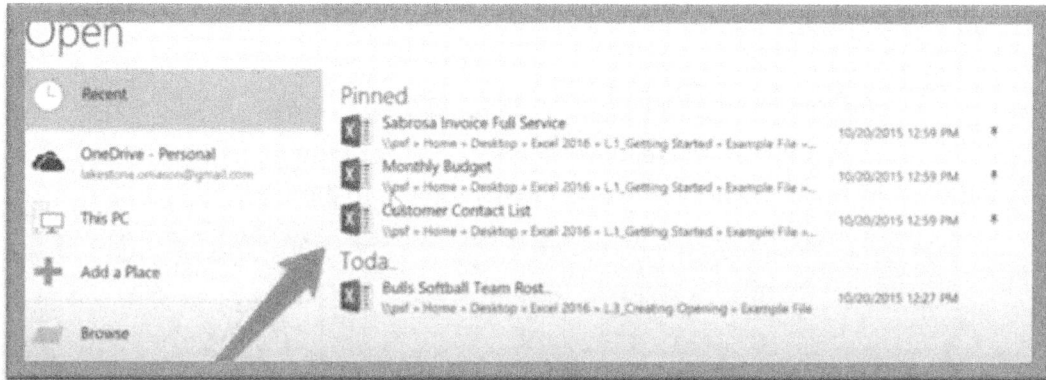

1. This opens the backstage view open page.
2. Choose the file you want to open by doing one of the following:
3. Select the file from the most recent list.
4. Select the file in a different location from the navigation list.
    5. Navigate to the desired file using the Browse button, and select the file from the Open dialog. Then select *Open*.

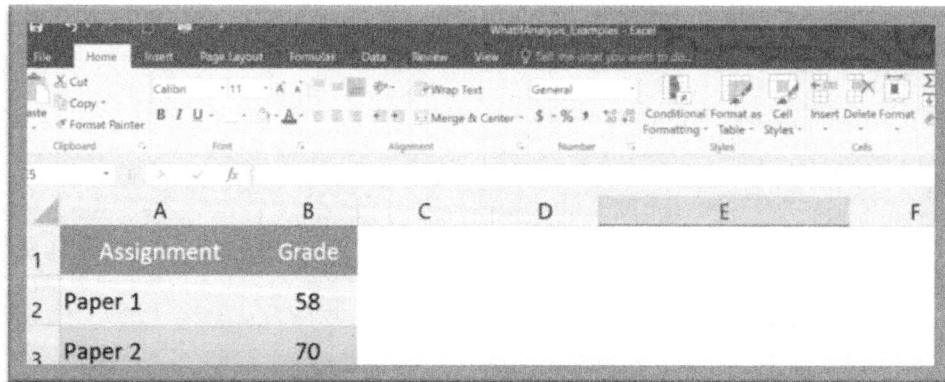

Working with Cells

Often when you type text in one cell, it extends into the nearby cell. You only want to have the text in the first cell, but it accidentally enters the second.

As a result, you must extend the cell containing the data so that you or the people with whom you share your Excel file can see all the text intended for that cell.

There are ways to make all of the text typed in a specific cell sit inside without cutting into other nearby cells. You can use the auto-fit option or manually adjust the cell width.

Let us start with the auto-fit option to extend a cell so that its content is displayed in that cell. To do this, click the cell that contains the text that extends to the one closest to it.

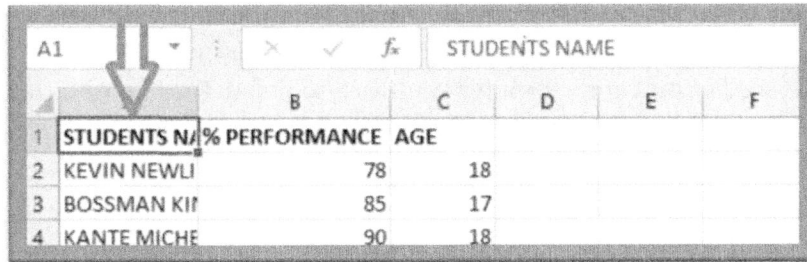

The Selected Cell with Data Indicated

In the above photo, cell A1 is selected because the full text is not displayed completely. The full text is *STUDENTS NAME, not STUDENTS N,* displayed in the photo.

Click the *Home* tab of the Excel spreadsheet

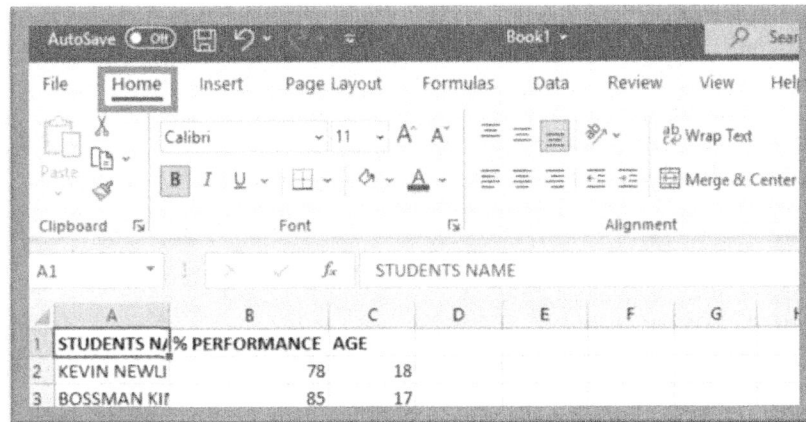

The *Home* tab Indicated

Under the Cells group, positioned at the top right-hand side, click *Format*.

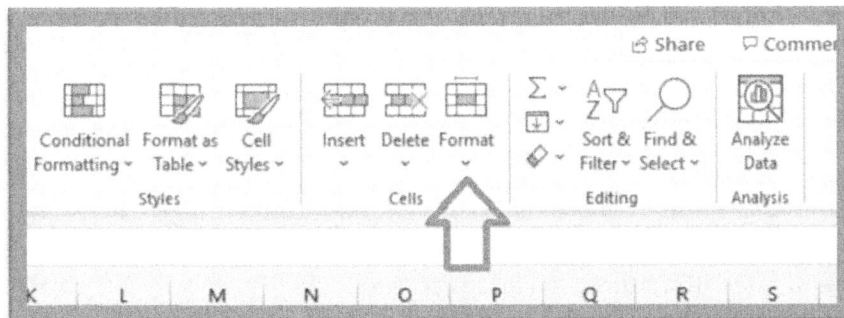

Progress in Extending Cell as Format Is Indicated

When you click Format, you will see some options. Just select *Auto-Fit Column Width.* Immediately after taking this last step, the cell is adjusted to show all text entered.

You should be aware that you can manually extend the cell as desired. To use this method, click the Format button in the Cells group and select Row Height or Column Width to make your changes. Enter the values you want for the row height or column width, then click the OK button. As soon as you do this, the cell expands. These are the details required to modify any spreadsheet cell.

How to Change the Text in a Cell

This is one area of spreadsheet use that many beginners struggle with. Instead of deleting a few letters from a word they previously entered in a cell, some people delete the entire contents of that cell. This occurs because many users are unsure of what to do.

If you only want to delete a few letters from a word you typed into a cell, double-click the cell containing the word you want to edit and then use the arrow keys on your keyboard to navigate to the letter you want to delete. Finally, press the Backspace key on your computer keyboard to delete the text or letter.

Furthermore, if you want to delete the entire word contained in a cell, click on the cell containing the word. Finally, press the Backspace key on your PC keyboard to delete the entire text in the cell.

Cells can also be formatted.

Excel is similar to Word in that there are many ways to structure and organize the contents of cells. Go to the Menu bar to access various editing features such as font, italics, underlining, and color options. Textured and colored cell backgrounds are both possible.

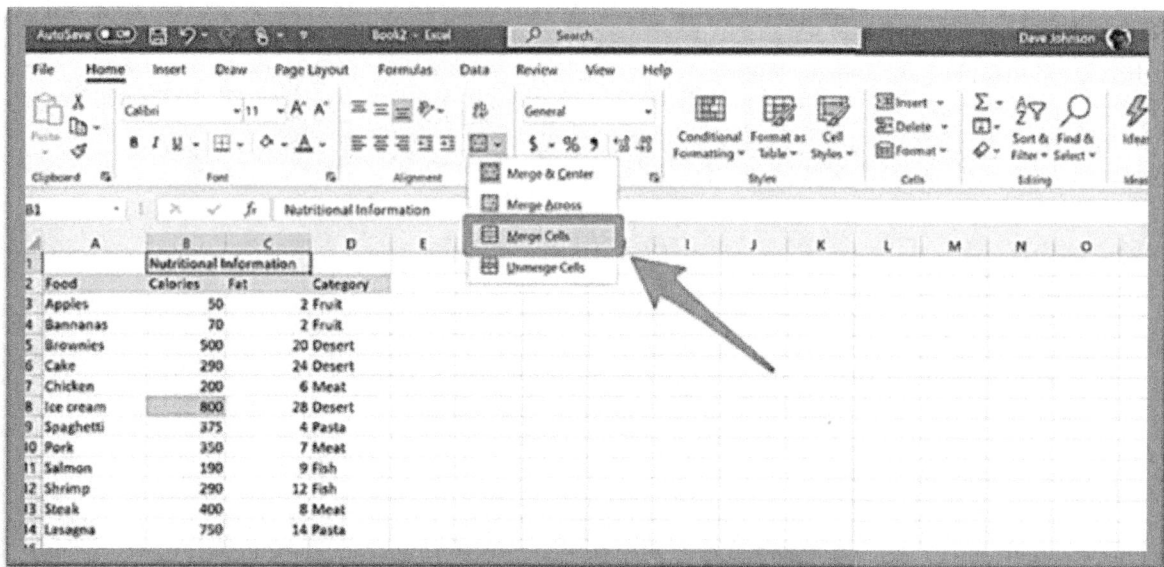

Remember that for spreadsheets, each cell is treated as an independent component. Because distinct numerals, characters, or phrases cannot be arranged individually within a cell, the bolded text and textual color will be the same. A cell's dimension can be changed by dragging its top border. If certain cells in a column have data that extends beyond the edge of the column and is hidden by data to the side, move the cursor over the separation seen in the middle cells in the column headers at the start of the worksheet. Then, select it to expand all columns.

A further option to ensure the content is visible is to reformat the cell so that the word wrap inside it instead of adjusting the cell's dimensions. You can also combine multiple cells into the same one, which would be useful for designing a column with both words and pictures.

Adding Comments in Cells

One of the advantages of using an Excel spreadsheet is that you can add comments to cells containing data to

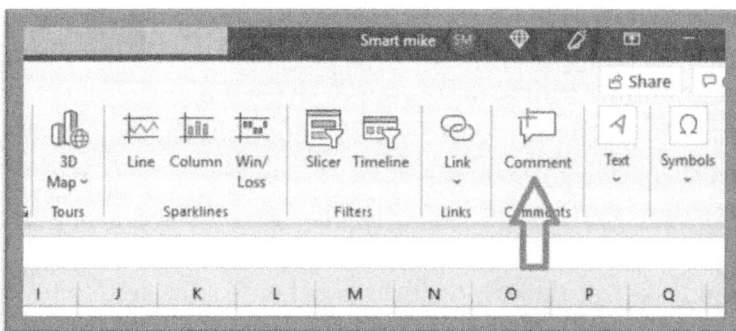

explain what is in that cell. Comments are added to cells to explain the data in that cell further. If you notice that the recipient of that Excel file may not understand the data in some spreadsheet cells, you should add comments to explain your points further.

For example, assume you have two employees in your company who share the same name but work in different departments. If you are adding the names of your employees and there are two employees with the same name, you could differentiate between the two by using comments. Perhaps you could include a note in each of the comments about the position that each person occupies.

The comment contents are displayed when the recipient moves the computer's cursor to the cell's comment symbol.

To add a comment in a cell, take these steps: Click the cell you want to add a comment in. On selecting the cell, click the *Insert* tab followed by *Comment,* which is positioned at the top-right of the interface.

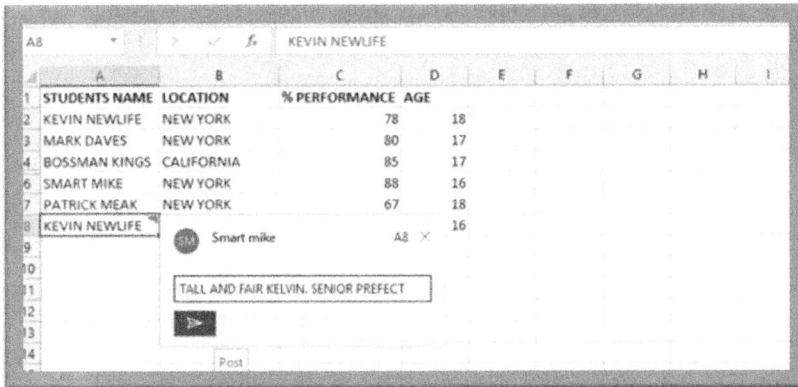

Figure 1Added Comment in a Cell Ready to Be Posted

The Comment Command Pointed by Arrow

The comment space will appear in the cell when clicking the Comment command. Start typing the note you want to add as a comment.

When you are done adding your comment, click the Post icon, which is an arrow pointing toward the right. Adding comments in a cell is a way to clarify some things. Once you click the Post icon, click out and continue with your other tasks.

Adding Notes in Cells

Notes do the same job as comments. Both are ways to add additional information in any cell to explain further to your recipient or yourself. So, whichever one you choose between notes and comments, they are all fine.

To add a note to any cell of your spreadsheet, first select the cell you want the note added. Remember that you can select a cell just by clicking on that particular cell.

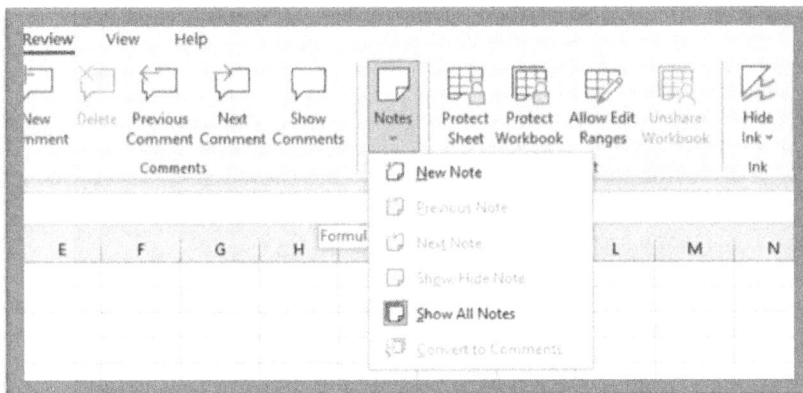

Figure 2Notes command selected

As the cell is selected, click the *Review* tab followed by the *Notes* command.

On clicking the Notes command, you will see some options. Just click *New Note.* Type the note in the space provided.

When you're finished typing your note, drag it to any part of your spreadsheet, so it doesn't distract you from reading the rest. When you add the note, you will see a red symbol in the cell. That indicates that there is an additional note in that cell.

In addition, when you add a note to a spreadsheet cell, the note will appear on the spreadsheet. Click the Review tab, then the Notes command to hide that note so it doesn't distract you from doing other things. Select the Show/Hide Note option. As a result, the added note will be hidden.

How to Get Rid of a Note

If you decide you don't need a note anymore, you can delete it. To delete a note, click on the cell that contains the note to select it. The next step is to right-click on the cell in question. Some options will be displayed. Simply select the Delete Note option. This deletes the note that was previously added to the cell.

Tips for moving and copying cells in Excel

1.   Dragging and dropping cells allow you to move them.

2.   Individual cells or ranges of cells can be moved or copied.

3.   Place the cursor at the edge of the selection.

4.   When the move pointer appears, drag the pointer to move the cell or range of cells.

5.   Cut and paste cells to move them.

6.   Make a selection of one or more cells.

7.   Click Home > Cut or press Ctrl + X.

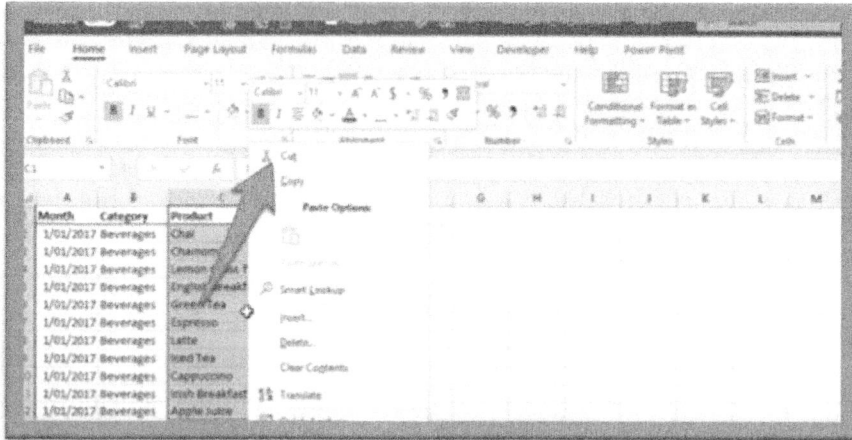

8.   Choose the cell where the data should be moved.

9.   Click Home > Paste or press Ctrl + V.

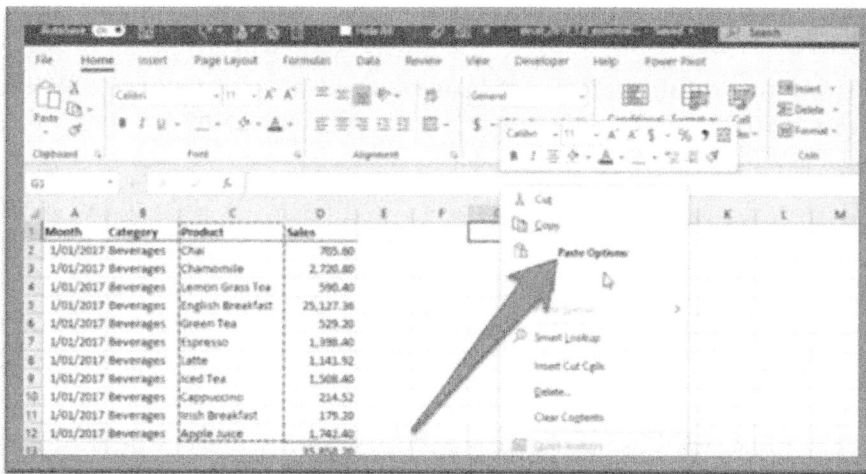

10.   You can copy and paste cells in your worksheet using the Copy and Paste commands.

11.   Cells or ranges of cells should be selected.

12.   Click "Copy" or press "Ctrl + C."

13.   Ctrl + V or Ctrl + V are shortcuts for Paste.

Copy or move just the Contents of the Cell

Click twice on the cell containing the data you want to move or copy.

Double-clicking the cell allows you to edit and select the data directly, but you can also edit and select it in the formula bar. Click on the characters you wish to copy or move in the cell.

Quick Access Toolbar Explained

The Quick Access Toolbar is built into all major Microsoft Office desktop applications. The Quick Access

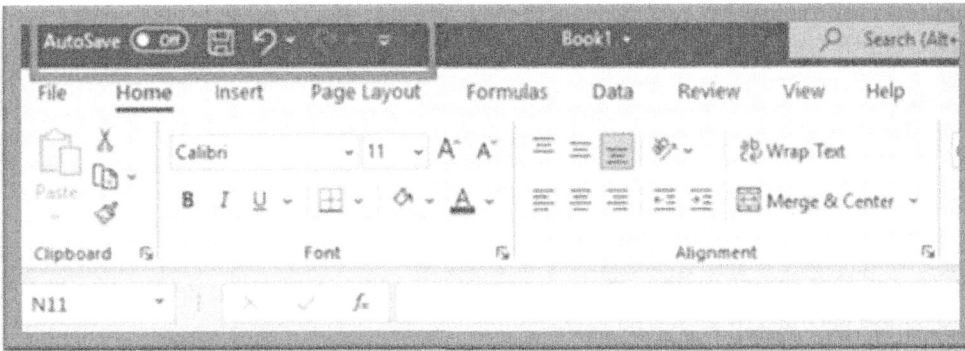

Toolbar includes tools or buttons that Excel users will likely interact with frequently. For those using the Excel 365 desktop app, the tools made available in the Quick Access Toolbar of the Microsoft Excel app by default are the Save icon, Undo, Redo, and AutoSave buttons.

These tools are standard and you can add some other tools to this as part of your Excel interface.

The Quick Access Toolbar section

In the above photo, the tools available in the Quick Access Toolbar section by default are rectangular. From left to right, they are the AutoSave button, Save, Undo, and Redo. These are used frequently by Excel users, which is why Microsoft positioned them there.

You can use the AutoSave button to save the Excel file you're working on in the Microsoft cloud app. OneDrive is the official name for Microsoft's cloud app. The account owner can access any file saved in OneDrive from anywhere worldwide. You can purchase any book from Amazon to learn more about the OneDrive app.
When you enable AutoSave, you will be prompted to sign into your OneDrive account. When you sign into the app, select the folder where you want to save the file. Before you can do this successfully, ensure your computer is connected to the internet. When you are successfully signed in, any changes you make to your Excel file will be automatically synced to OneDrive. The AutoSave icon is enabled in this case.

Another tool is accessible via the Save icon. When you move your computer cursor over it, it will say "Save." To avoid data loss, use the button to save the information you entered in your spreadsheet. As you enter data into a spreadsheet, frequently click the Save icon to save your work. More importantly, click the Save icon before closing that Excel spreadsheet file. This will save the last changes you made to the file before closing it.

The *Undo* button is another tool visible in the Quick Access Toolbar section in the above image. You can undo your previous action on the Excel spreadsheet by clicking the Undo button. For example, if you accidentally entered someone's name into the spreadsheet, you can undo that action by clicking the Undo button.

The *Redo* button is used to redo something you have undone. After you have undone something, you can click the Redo button, which takes you to the place you were previously. It gives you an immediate result without taking any time.

File Sharing in Excel

Click the Share button at the top right to share your spreadsheet file after its completion.

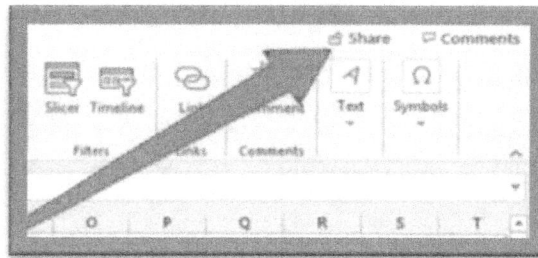

The Share button pointed by the arrow

On clicking the Share button, Excel will request how you want to share the file. You can share the file on OneDrive, share as attached in workbook format, or as a PDF.

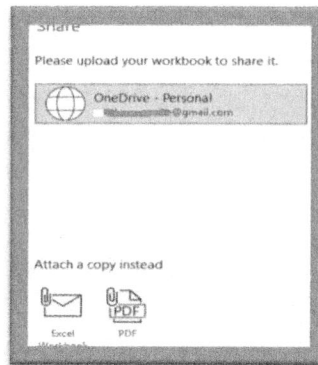

Options available to share Microsoft Excel file

Out of the 3 options, select the way you want to share your file.
Sharing Your Excel File on OneDrive.
To share your Excel file on OneDrive, click the share icon at the top right corner of your Excel spreadsheet.

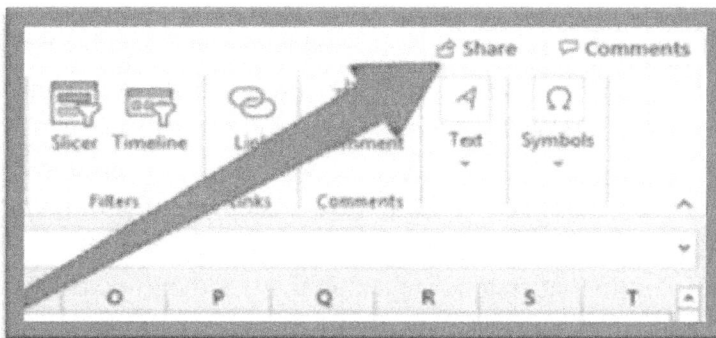

The Share button pointed by arrow

On clicking the Share button, select the *OneDrive* option. If you are not signed into your OneDrive account automatically, you must type in your email address and password to sign in. Choose the folder in which you want to save the file and then save it there.

Sharing Excel File as Attach

As mentioned above clicking the Share button at the top right of your Excel spreadsheet interface allows you to share the file as an attachment. You can attach a copy in a spreadsheet or PDF format when you select the option to attach a copy. If you want to attach a copy of the spreadsheet, select Excel Workbook from the Attach a copy menu instead. To log into your email, simply follow the prompt. There will be space for you to type the email addresses of the people with whom you want to share the file. Fill in their email addresses. When you're finished, click the send button to share the Excel file with them.

These are the steps you need to take to share Excel with your OneDrive account and with other recipients.

Chapter #4: Data input

Entering Data

The first step in creating the workbook is manually inserting data into your worksheet. The stages below demonstrate how to type column headings throughout the document. In row 2 of a worksheet:

1. Go to cell A2 on your worksheet and click.
2. Enter a month's name, for example.
3. Move the cursor to the right using the right cursor keys. Cell A2 will be activated after the term is entered.
4. To create unit sales, use the right cursor key.
5. Repeat step 4 with the terms average price and sales in dollars.

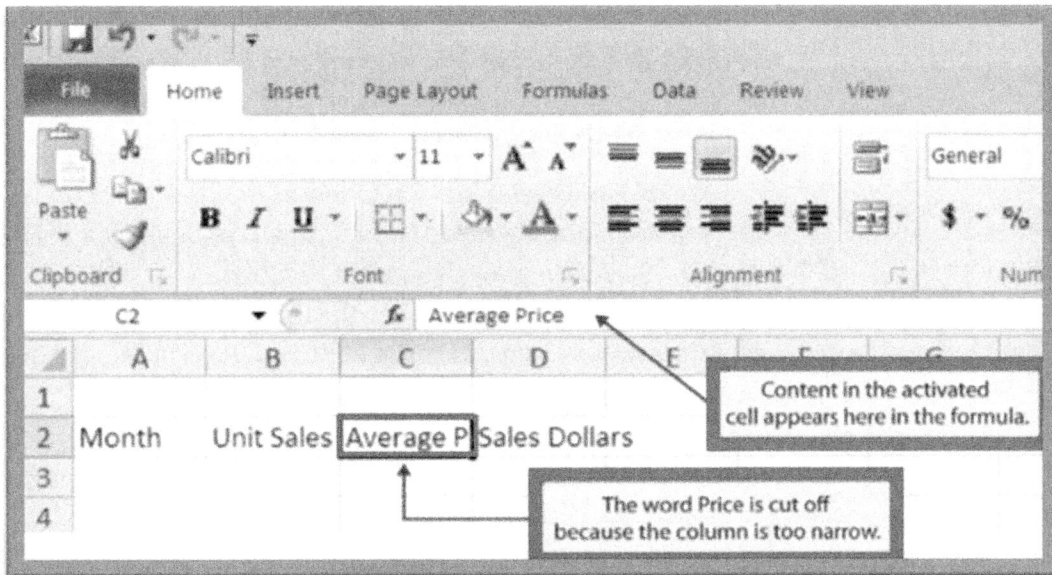

Once you've typed your column headings in row 2, check figure 1 and see how the worksheet looks. It's worth noting that the term price in cell C2 isn't available. This is because the columns are too narrow to accommodate your typed entry.

While entering numbers, don't use symbols to format them. It's better not to use coding symbols like dollar signs and commas while entering numbers into an Excel worksheet. While you can incorporate these symbols when typing numbers in Excel, it delays the data entry operation. It's easier to apply these symbols to numbers after they've been typed into the worksheet using the formatting features of Excel.

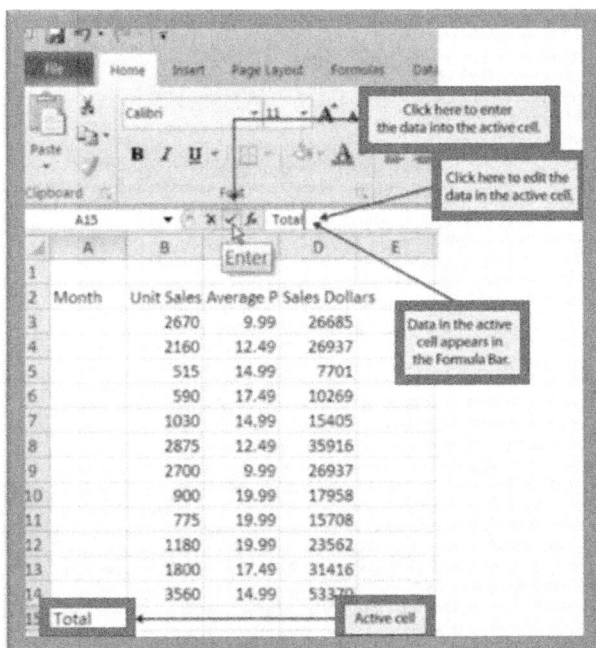

Editing Data

Double-clicking a cell position in the formula bar will change the data. You may have noticed that the data was typed into the cell position when you typed it, as shown in the formula bar. The formula bar can insert data into cells and modify previously entered data. The following stages demonstrate.

How to enter and then edit data entered into a cell position:

1. Click on cell A15 on sheet 1 of your worksheet.
2. After typing the abbreviation 'Tot,' press the enter key.
3. Cell A15 should be chosen.
4. Raise your mouse pointer to the formula bar. The pointer will become the cursor. After moving the mouse to the last letter, left-click on the abbreviation 'Tot.'
5. Type the letters 'Al' to finish the term total.
6. To the left of the formula bar, click the check mark. The change is then made in the cell.
7. Cell A15 should be double-clicked.
8. Enter the word sales after the total, with a space between the two words.
9. To begin, press the enter key.

Moving Data

1. To move content, you follow a similar set of actions as we did with copying; however, you would *Cut* the data instead of *Copy* it.
2. Select the range you want to move.
3. On the *Home* tab, click the *Cut* button (this is the command with the scissors icon), or press *CTRL+X*. A scrolling marquee will appear around the area you've chosen to cut.

4. Place your cursor on the first cell of the area where you want to paste the content. You only need to select one cell.
5. Click on *Paste* on your toolbar. This will move the content from its current location and place it in your chosen area.
6. The cut-and-paste action automatically copies the cells' format across but not the width. So, you need to adjust the width of the cells if necessary.

Copying and Pasting Data

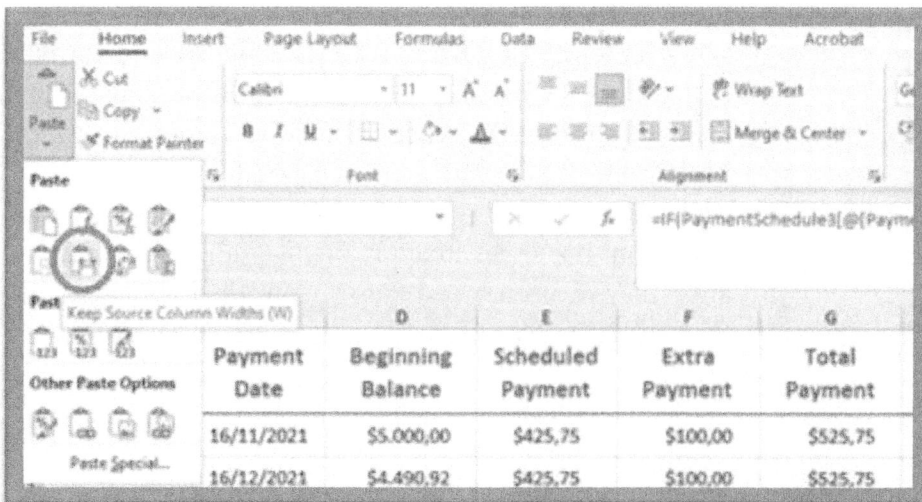

Simple copy and paste:

1. Click Copy on the Home tab, Clipboard group (this is the double paper icon next to the Paste command).
2. Click Paste in the first cell of the area where you want to paste the contents.
3. To speed up the process, use the same procedure but replace the Copy button with CTRL+C and the Paste button with CTRL+V.

4. The marquee remains active to inform you that you can continue pasting the copied content if you want to paste it in multiple places.

5. To remove the marquee, press the ESC key.

Other options for pasting:

1. After you've copied data, click the Paste command button on the toolbar to bring up a pop-up menu with several pasting options.

2. You can hover your mouse over the options to see what they do.

3. On your worksheet, you can also see a preview of the paste action.

4. For example, if you want to paste the contents as well as the column width, you would select the Keep Source Width option (W).

5. This is on the second row of the menu.

6. Select that option to paste the data as well as the cell formatting and column width.

7. Once done, remove the marquee around the copied range by hitting the *ESC* key. This tells Excel that you're done with the copying.

Drag

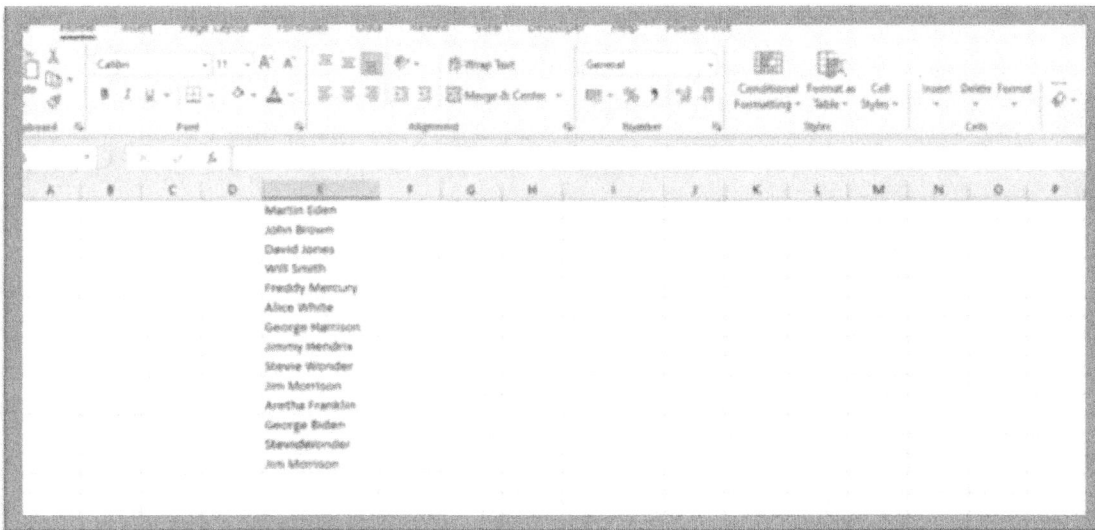

Now we will talk about another important Excel function: dragging. It is simple, immediate, very fast, and will allow you to create ranges of cells. You will need this a lot when creating automatic lists. How do you drag? You do it by moving the cursor at the base of the cell. You can do it vertically, horizontally, and even diagonally. We will soon see how useful this feature can be.

Remove Duplicates

In larger datasets, duplicate material is common. Someone may have a variety of connections in an organization and want to know how many they have. In situations like these, removing duplicates is critical.

To remove duplicates, select the row or column from which they should be removed. Then, on the data tab, click "Remove duplicates" (underneath tools). A pop-up window would appear, prompting you to confirm which details you intend to use. Simply select "Remove duplicates" and you're done.

This function can also delete an entire row based on the column's duplicate values. If someone has three rows of information about a movie and only needs to see one, they can select all datasets and then delete duplicates. The final list can only include specific titles, with no duplicates.

Excel Lists

In the Excel settings, you already have some sorting of numerical, alphanumeric, or year, month, day of the week, and more. These lists are set by default in the program, but you can set up extra ones that may serve you in your work.

Let's try to do it together. The first thing to do is to continue working on our worksheet by adding other data to the list of names, such as the customer number, day, month, and year, and you will find that with the automatic series it is very easy, especially if you find yourself managing lists with 1000 customers or 1000 suppliers. Now, you have to get this result with a few clicks. How?

As you can see from the image, you have to pull down the marker and the number 1 will be repeated, but clicking on the highlighted arrow will open a drop-down menu, there you will make your choice by clicking on copy series, that's it, the numbers will come in succession.

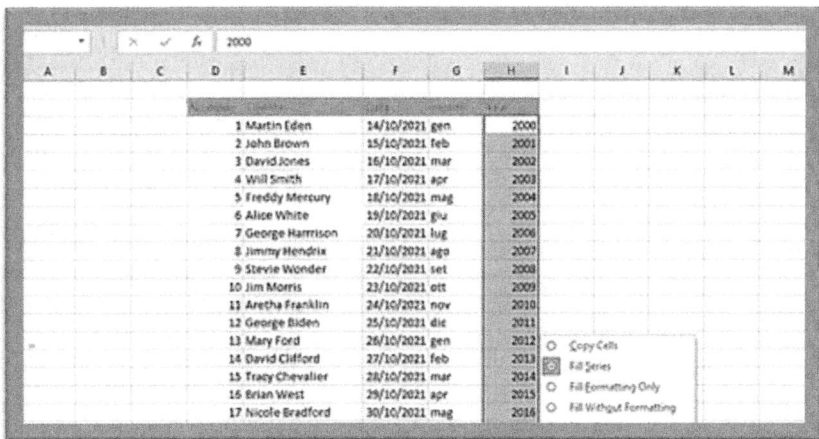

This also applies to dates, such as days of the week or months of the year. These are pre-programmed series because they are the most popular. However, each of us may require different settings, as well as different types of series. In this case, let's look at where Excel has set up the series so that we can add what is useful for our daily work. Go to File and then to the bottom left, you'll see Options. Click on it, and a settings window will appear; go to Advanced and select the option shown in the image. When you click on Edit Custom Lists, a window opens in which you can see all of the default lists and add anything you might need for your work, as shown in the image.

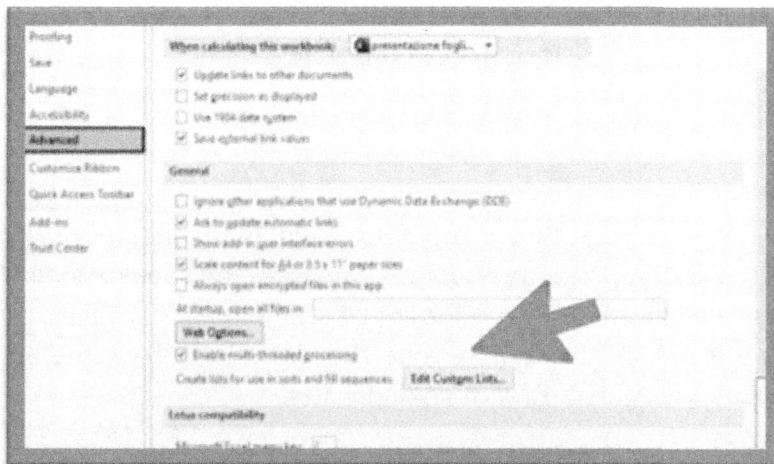

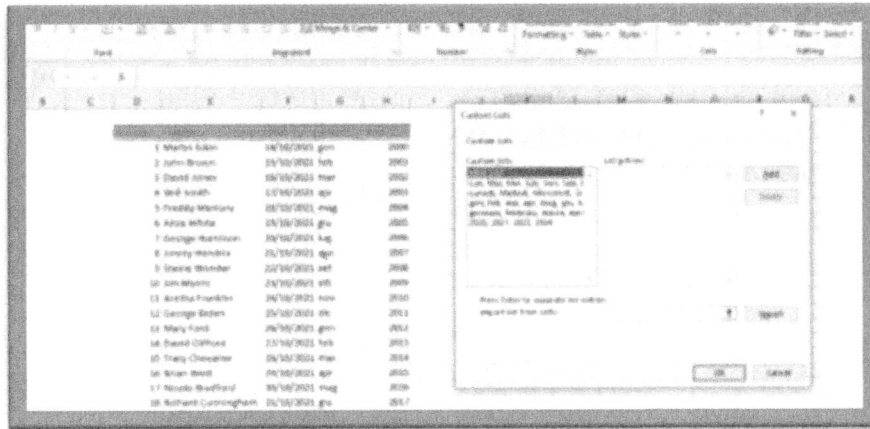

Slowly, we have outlined an overview of the basic functions, and above all, we begin to understand that Excel is not the bugbear it first seemed. All you need is a little practice and you become masters of the tool that really simplifies life. Just think of a customer or supplier list as you manage it quickly and above all with precision and without errors.

Format Painter

Excel has a range of characteristics that render analyzing numbers and analyzing data fast and simple. However, if one has ever spent time customizing a sheet, they know how boring it can be.

Don't spend time repeatedly entering the same formatting instructions. If you want to quickly copy the formatting from one part of the worksheet to another, simply use the Format Painter tool. To use it, select a cell one wants to duplicate, then go to the top toolbar and select the format painter choice.

Deleting Data and the Undo Command

The most common way to remove data from a computer is to use the delete key or the clear button on the ribbon. You can also delete rows, columns, or cells to remove data.

Let's take a look.

1.   One method for removing Excel data is using the Clear button on the home ribbon.
2.   If you only want to remove data, choose "Clear Contents."
3.   Select "Clear All" to remove both the contents and the formatting.

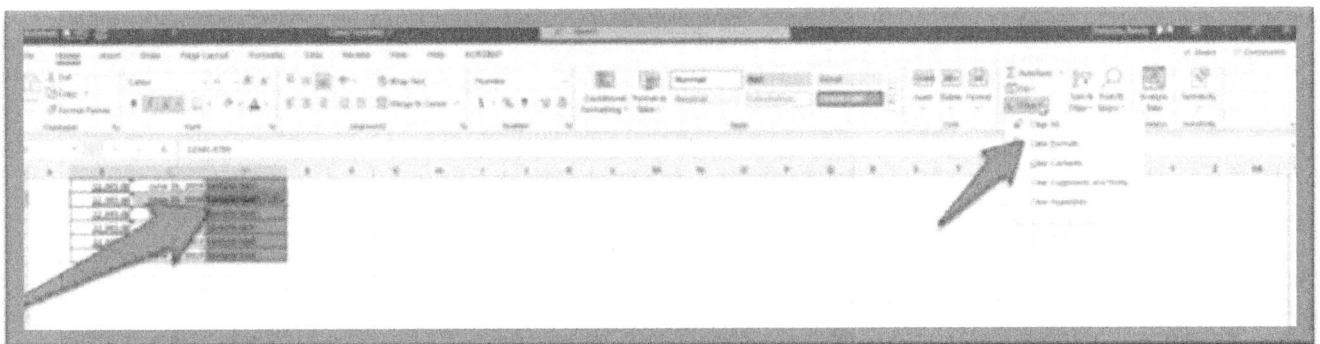

4.   The delete key is a faster way to clear content. You can delete the cells by selecting them and then pressing the delete key.
5.   Deleting cells in this manner removes the data rather than the formatting. If you want to remove the formatting as well, use "Clear all" in the Clear section of the Home ribbon.

You can also delete entire rows or columns to remove data from a worksheet.

1.	Select the rows or columns containing the data you want to delete and delete them using one of the methods described previously.
2.	You can use this method to quickly clean up a worksheet by removing all data and formatting.
3.	You can also right-click and select "Delete" from the context menu.
4.	When the Delete dialog box appears, choose an appropriate option.
5.	When removing rows or columns, it is advisable to consider any other data that may be present in other parts of the worksheet.

If you have entered data in error and don't want to go through the stress of having to keep pressing the backspace or delete key, you can press Ctrl+Z. This will undo whatever mistake that may have been made.

Find and Replace

Find and Replace is a useful feature of Excel that is often overlooked.

Find and Replace Text and Numbers in Excel

Looking for a specific value in a large spreadsheet is a common task. Fortunately, this is made easy using Find and Replace.

1.	Click any cell or choose the column or range of cells you want to analyze.
2.	Select Home > Find & Select > Find or press the Ctrl + F keyboard shortcut.
3.	Enter the text or number you are looking for in the "Find What" text box.

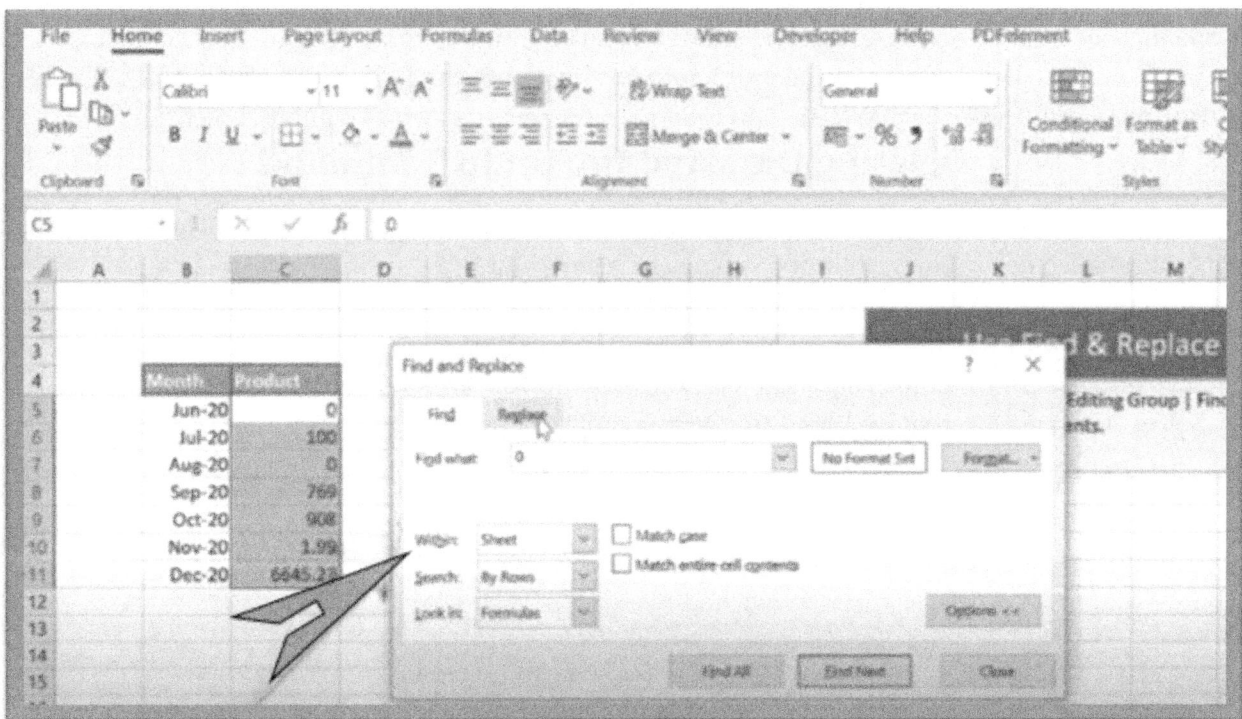

4.	Click on "Find Next" to find the first instance of a value in the search area; click "Find Next" again to find the second instance, and so on.
5.	When you select "Find All," you will see all instances of the value and information such as the book, sheet, and cell where it appears.

To go to a specific cell, click on it.

1.	To save time browsing through spreadsheets, finding specific or all instances of a value can be useful.
2.	Select "Replace" from the menu for any occurrences of a value you want to replace.

3.   Enter the text you wish to replace or the number within the "Replace With" text box.

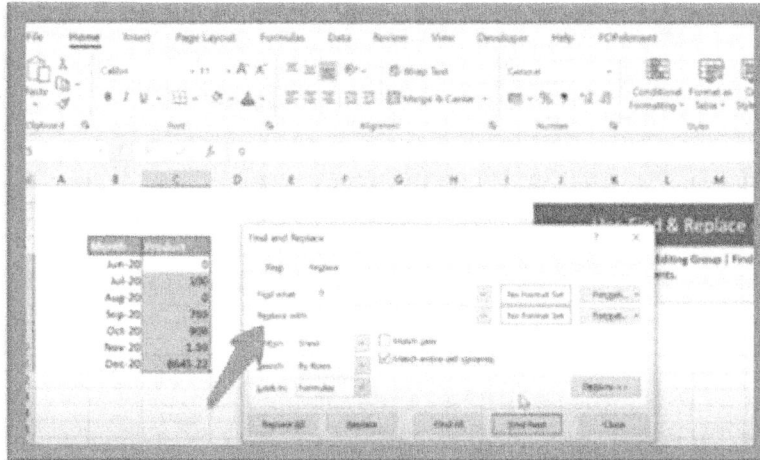

4.   Click "Replace" to modify each occurrence, or choose "Replace All" to modify all occurrences simultaneously.

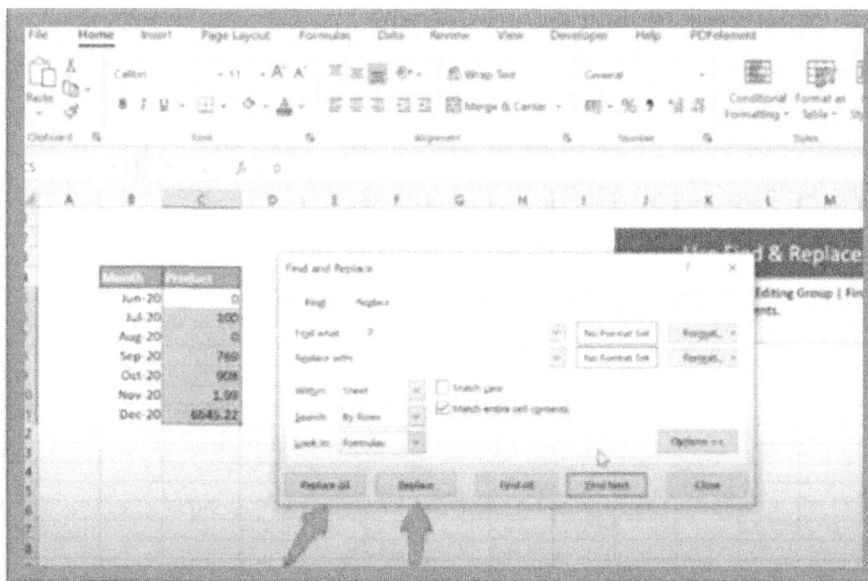

Values Should Be Formatted Differently

Value formatting can also be found and replaced.

You can search for and replace a range of cells or click on any cell on the worksheet to search the entire active worksheet.

5.   To open the Find and Replace window, go to Home > Find and Replace and then click on Replace.

6.   To view the Find and Replace options, click the "Options" button.

7.   You do not need to enter text or numbers to find and replace them.

8.   To format the "Find What" and "Replace With" fields, click the "Format" button next to them.

9.   If you want to replace or find a specific format, specify it. The Find and Replace window displays a preview of the actual format. A preview of the formatting is displayed

You may also set any additional options you wish, then click "Replace All" to change all the locations where the formatting is used.

Chapter #5: Formatting

Table Formatting

The user can format Excel using the tools available in the Font group of the Ribbon. Font styles, fill colors, and borders can all be customized. You must be careful not to move cells around too much when creating the format. It is very easy to make mistakes with the border format or the fill color. If you want an organized structure with consistent colors, you will love Tables.

Making a Table Structure

1.   Return to Cell A1 and repeat the procedure (Ctrl Home).
2.   Choose Format as Table 3 from the Home tab, which is next to the Conditional Formatting option. Select a color scheme that alternates the colors in each row.
3.   Excel should load the entire dataset. Because we already have titles and headers, that option will be unchecked. Click the OK button to see the result.
4.   For the DUE DATE section, the conditional formatting will still be in effect.

We have a tab on the ribbon that will allow us to change the table's design.

Experiment with different table styles and layouts to see how they affect the overall format of the table. One of the most useful features is the Total Row.

After you have enabled the entire row option, scroll to the bottom of the data. The number 77 reflects the number of records we have. Change the Total for the BALANCE row to Sum by clicking within the Total for the Balance column.

Excel Templates

Excel provides you with premade templates, or a user can make their own. Doing so will require you to invest some time, but it will save you a lot of time later on.

•   Exploring Excel template: The easiest way to familiarize you with the Excel sample files is to hop in and check them out.

•   Viewing templates: Click on File and then New to explore the templates. The prototype thumbnails on the screen are only a small selection of the available ones. Select one of your search words listed, or type in a specific term and check for more.

Enter the invoice, for instance, and press the Search button. Several thumbnails are shown in Excel. You may use the group filters on the right to narrow down the results. Below is a sample invoice template.

How to Use Them

To create a template-based workbook, locate a prototype that appears capable of doing the job and tap on the thumbnail. Excel displays a box containing a larger version, the template source, and some additional material. If everything still looks good, click the Build button. If not, use your arrows to view information for the list's next (or previous) design.

When you click the Create button, Excel will import the prototype and create a new workbook based on that design. The operation you perform may differ depending on the template. Although the prototype is specific, much of it is self-explanatory. Tailoring is covered in many workbooks. Replace the default details with your own.

The image below depicts a workbook created from a blueprint. For the user's needs part of the workbook must be customized. Customisation is a way to improve the templates automatically generated.

The one below is a workbook created from a template.

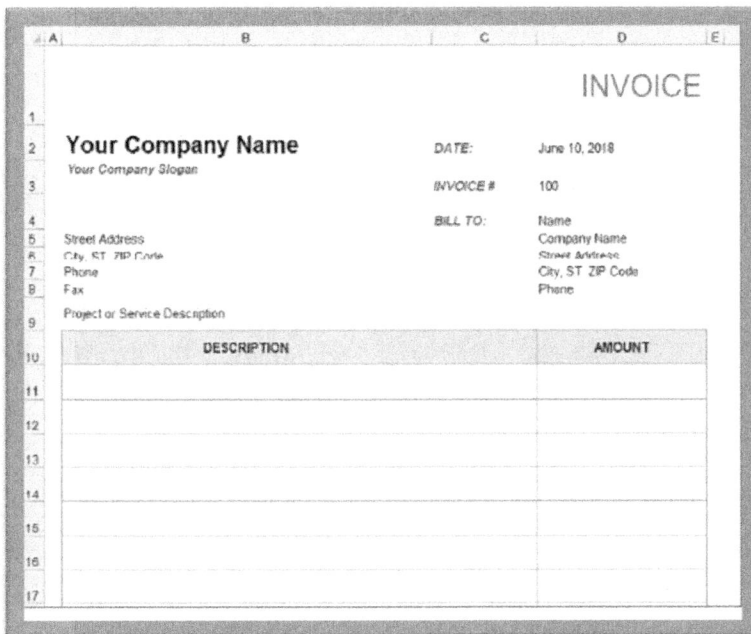

Use the save command for later use of this document. Excel provides a filename based on the prototype title, so you can choose whatever filename you choose.

• Using Default Templates: Excel provides three types of using the file. The original template for the workbook. This type is being used as the base for new workbooks. The default template is used as a base for later use while working with other files.

These premade template files include formats that can be used while working with other files. It will help you to save time and work effortlessly.

Number Formats

When working on a spreadsheet, you should utilize appropriate numeric formats for your data. Number formats define the details you'll use in your spreadsheet, such as currency ($), percentages (%), dates, times, zip code, phone number, and so on. Number formats improve the readability and usability of the spreadsheet. When you add a number format to a cell, you inform your spreadsheet about the type of values stored there. For example, the format of the dates indicates to the spreadsheet that you are inputting particular calendar dates. This enables the spreadsheet to effectively represent your data, ensuring that your data is accurate and your formulas are correctly measured. The spreadsheet will likely use the general number format by default if no number format is specified. In contrast, the general format can make minor formatting changes to your data.

Applying number formats

Number formats are used the same way that other types of formatting are, such as changing the font color, by selecting cells and selecting the desired formatting option. There are two ways to choose a number format. Select the required format from the Number group's Number Format dropdown menu on the Home tab.

One of the convenient number-formatting commands is displayed below the dropdown menu.

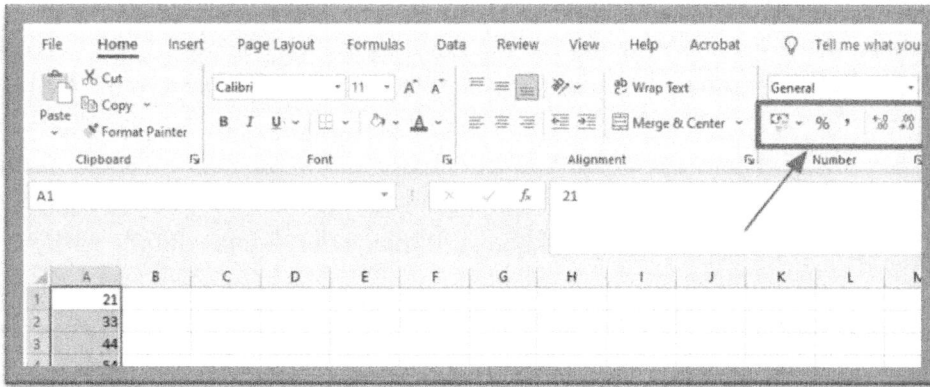

Select the required cells and press CTRL+1 on your keyboard to access more number-formatting options.

In this example, we've used the Currency number format, which introduces currency symbols ($) and shows two decimal places for any quantitative values.

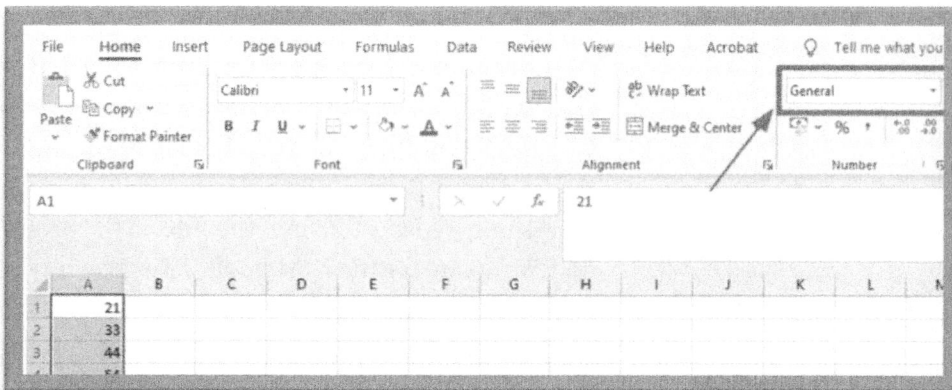

If you choose any cells with number formatting, you will see the actual value in the formula bar. This value will be used in formulas and other calculations in the spreadsheet.

On the *Home* tab, locate the Number group and click the dropdown list to display a number of formats.

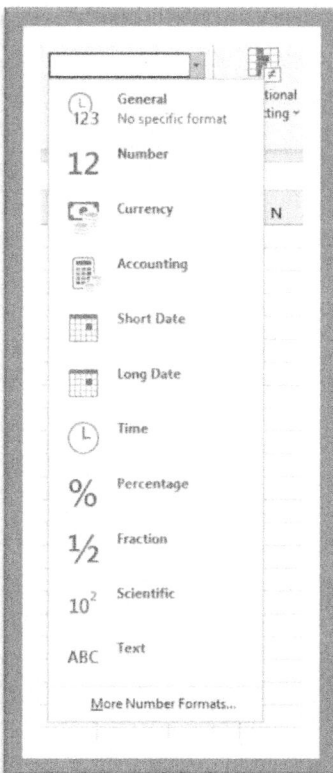

The selected cell/range will now be formatted in your selected format.

Accessing More Number Formats... allows you to select a format that is not available in the dropdown list; let's look at how to use it. For example, if you are in the United States and want to convert US dollars to British pounds. At the bottom of the dropdown list, select More Number Formats... (as shown above) or click on the dialog box launcher (the small diagonal arrow at the bottom-right of the Number group).

The Format Cells window will appear.

To display a dropdown list, click on the Symbol field. Choose the British pound sign (£) from the drop-down menu.

You can also specify the number of decimal places and the format for negative numbers on this screen.

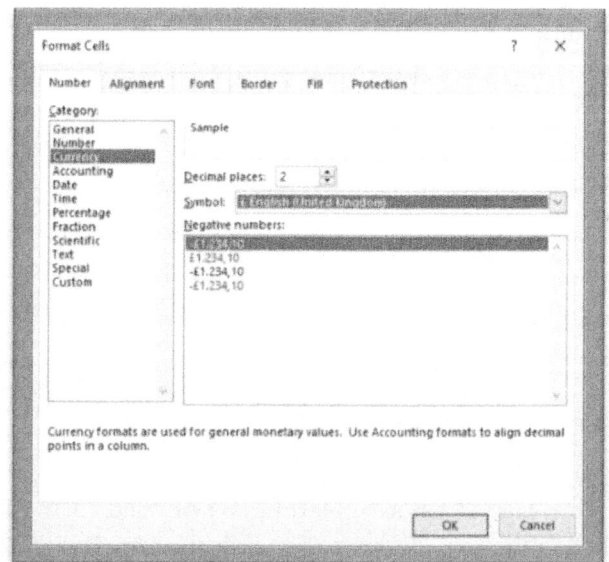

The Sample field previews how the selected format will appear on your worksheet. Click OK to confirm your changes when done.

Creating Custom Numeric Formats

If none of the existing formats fit your needs, you can design your unique format.

Assume you have a column in your worksheet where you keep track of a set of numbers. For example it could be unique product IDs, product serial numbers, or phone numbers. You may prefer that the numbers be displayed in a certain format regardless of how they were entered.

In Excel, you can define your format for a set of cells so that every data is formatted with your preferred format. To create your format: Right-click any area in your selection and choose *Format Cells* from the pop-up menu. Alternatively, launch the Format Cells window by clicking the dialog box launcher in the Number group on the Home tab.

Under Category, select Custom.

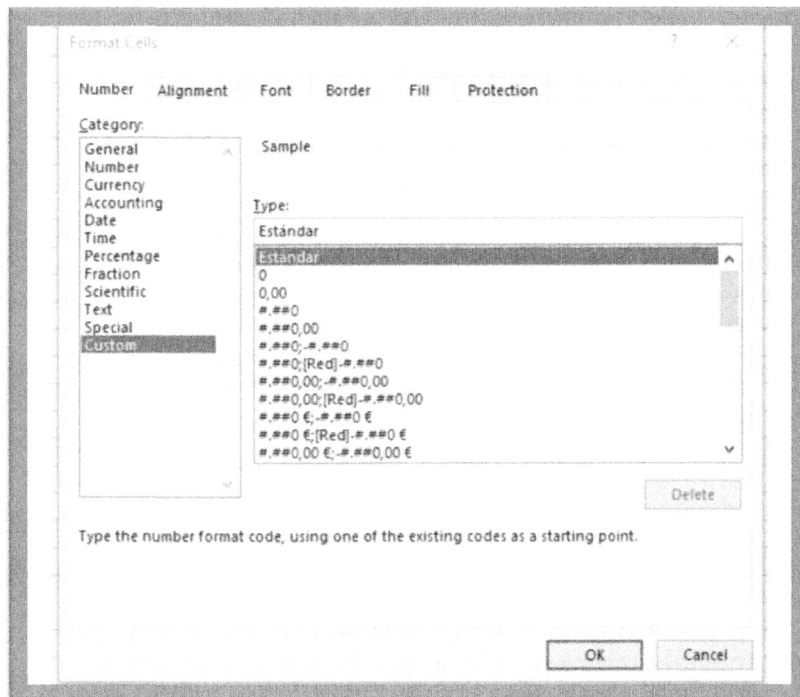

In the Type box, select an existing format close to the one you would like to create.
Note: If you find a format on the list that meets your needs, you can select that one and click OK.
In the Type box, type in the format you want to create. For example, 0000-00000.
Click OK.

In the image to the right, column A has a set of numbers. Column B shows the same numbers with a custom format (0000-0000000) now applied to them.

	A	B
1	Serial number	
2	77993871654	7799-3871654
3	77998976543	7799-8976543
4	77995678978	7799-5678978
5		

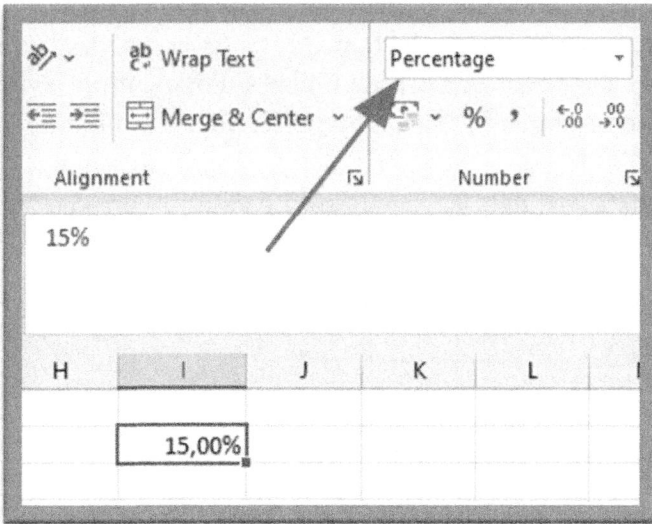

Percentage Formats

The percentage (%) format is one of the most useful number formats. Values are shown as percentages, such as 15% or 65%. This is particularly useful when estimating data like sales tax or a tip. The percentage number format automatically applies to the cell when you enter a number followed by a percent sign (%).

Percentage formatting can be helpful in different situations. As an illustration, observe the distinct configuration of the sales tax rate in each of the spreadsheets shown below (5.5%, and 0.05):

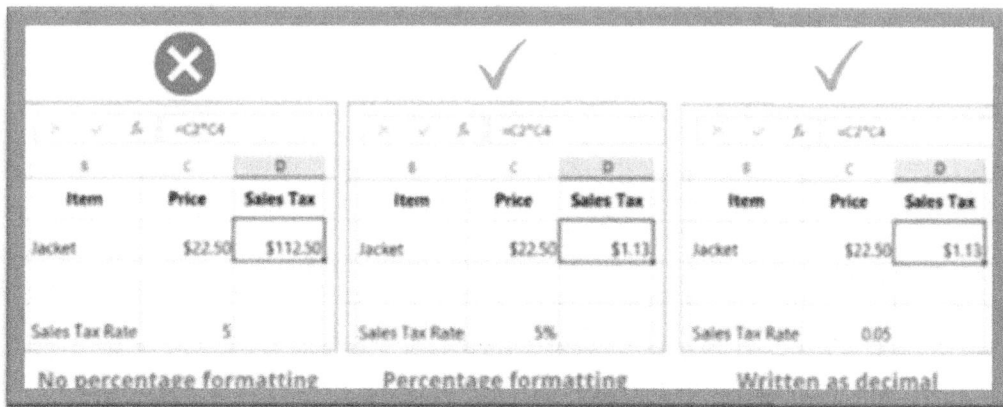

The calculation in the spreadsheet on the left appears to be incorrect, as you can see. Our spreadsheet assumes that we want to multiply $22.50 by 5 instead of 5% if we do not utilize the percentage number format. Although the spreadsheet on the right still functions without the percentage format, the spreadsheet in the center is more understandable.

Formats for Dates

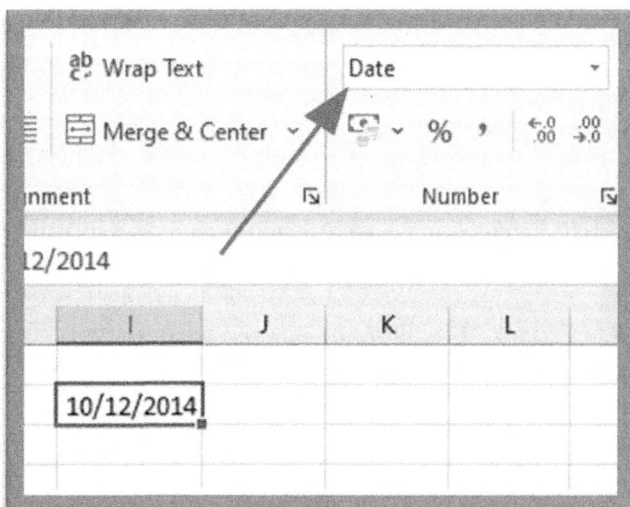

When dealing with dates, use a date format to indicate that you're talking about specific calendar dates, such as June 25, 2014. Date formats also provide access to a useful set of date functions to compute an answer based on time and date information.

Spreadsheets do not process data in the same way that humans do. If you type July into a cell, for example, the spreadsheet will not recognize it as a date and will treat it as any other text. As a result, you must enter dates in a format your spreadsheet recognizes, such as day/month/month/month/day/year. In the example below, we'll type 10/12/2014 for October 12, 2014. Our spreadsheet will automatically implement the date number format for the cell.

Now that we've formatted our data correctly, we can use it for a variety of purposes.

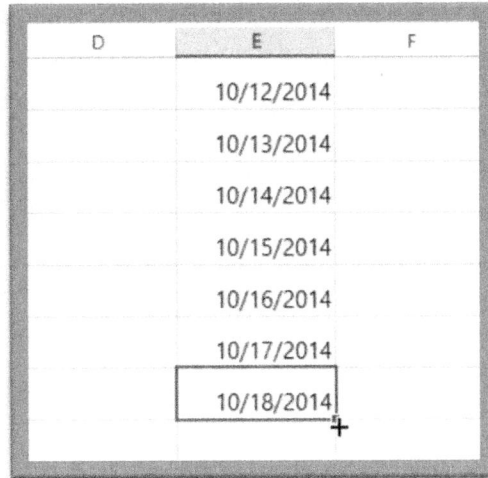

Since the spreadsheet didn't understand that we were referring to a date, this cell is still formatted as a number.

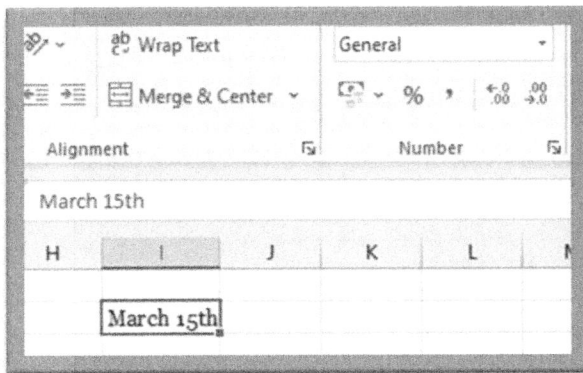

When we type March 15 (without the "th"), the spreadsheet acknowledges it as a date. Due to a year's absence, the spreadsheet would automatically add the current year, ensuring that the data contains all the required details. The date could be typed in various forms, like 3/15/2014, 3/15, or March 15, 2014, and the spreadsheet will acknowledge it as a date.

To see if the date format is automatically implemented, enter the following dates into a spreadsheet.

12th of October, 2014

10/6/2015 \s10/11 \sOctober 14th of October 12th of October

Alternative Date Formatting Options

Select More Number Formats from the Number Format dropdown menu to see more date formatting options. There are several options for displaying the date, such as including the day or ignoring the year.

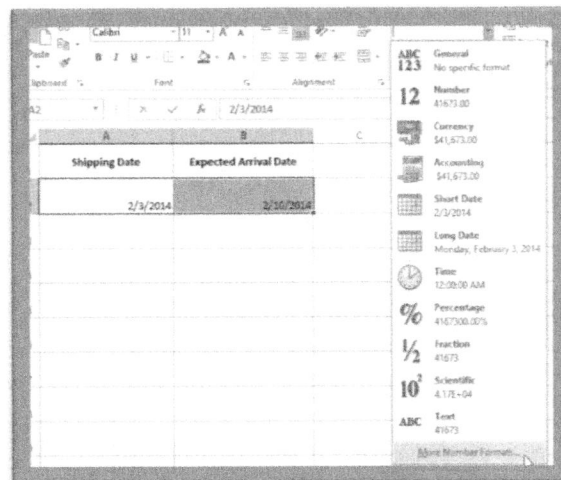

A dialog box called Format Cells will appear.

You can select the required date formatting option from this menu.

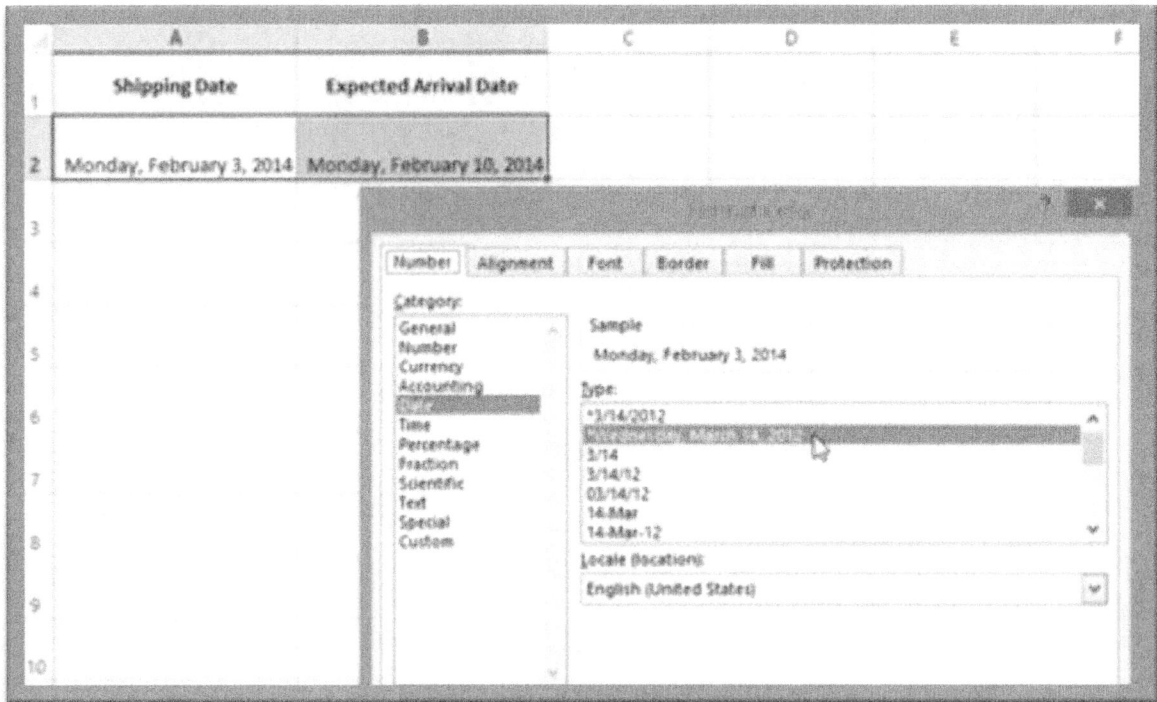

A custom date format doesn't change the date in our cell; it only changes how it's displayed, as you can see in the formula bar.

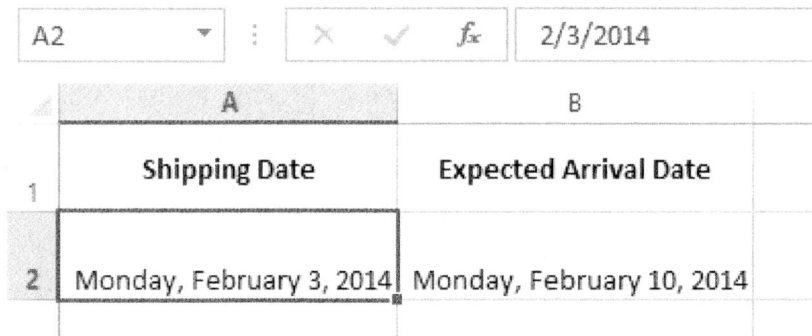

Copy Cell Formatting

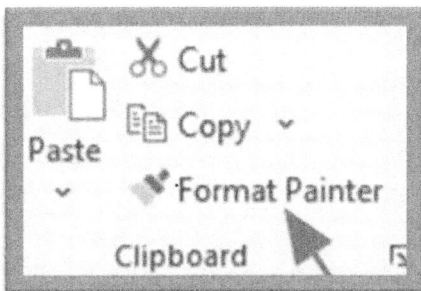

A quick way to format a cell or group of cells based on another cell is to use the *Format Painter*. This can be found in the *Clipboard* group on the *Home* tab. This can save a lot of time as you only create the format once and copy it to other cells in your worksheet for which you would like to apply that format.

To copy cell formatting with the Format Painter:

1.   Click on the source cell; that is the cell you want to copy the format from.

2.   Select *Home* and click the *Format Painter*.

3.   Click and drag over the destination cells, i.e., the cells you want to copy the format to. The destination cells will now have the same format as the source cell.

Example: If cell A2 is formatted as *Currency* and you want to format A3 to A14 as currency with the *Format Painter*, you would carry out the following steps:

4.   Click on cell A2 to select it.

5. Click on Format Painter.
6. Select A3 to A14, and the job is done.
7. The currency format from A2 will now be applied to A3:A14.

Add and Remove Columns/Cells

1. On the overview worksheet, arrange each category into a table with a header that corresponds to the grouping.
2. Select Format as Table again from the home menu and click on column A3 (Items) (Ctrl-T).
3. Select a table style that corresponds to the year.
4. To accept the range, click OK.
5. Repeat steps 3 and 4 for cells F3 (Items) & K3 (Items).
6. Select cell C7 and enter the letter D.
7. Hit Enter to advance to cell C8, where you should type E and press Enter.
8. Right-click on Cell F6 and choose to Add a row below from the context menu.
9. Repeat the process once more.
10. Fill in the blanks with B and D in the relevant places.
11. Find the little blue box with a backward L in the bottom right corner of N6 and press it.
12. Move the box back to two rows by dragging it.
13. Fill in the blanks with D and E in the relevant places.
14. In Row 7, observe that you cannot Insert or Delete text by right-clicking on the header of Row 7. These must be completed from the interior of each table.
1. Choose Cells C7 and C8 from the drop-down menu.
2. Right-click — Remove the rows from the table
3. Repeat the process for each table.
4. F7 and F8 key
5. K7 and K8
6. Alternatively, undo all of the newly inserted rows.

Text and Cell Alignments

There are three horizontal alignment options: against the left, against the right, and in the center. There are three vertical alignment options as well: against the top border, against the bottom border, and in the center. To modify the text alignment in Excel, select the cell(s) to be realigned, navigate to the *Home* tab > Alignment group, and pick the appropriate option:

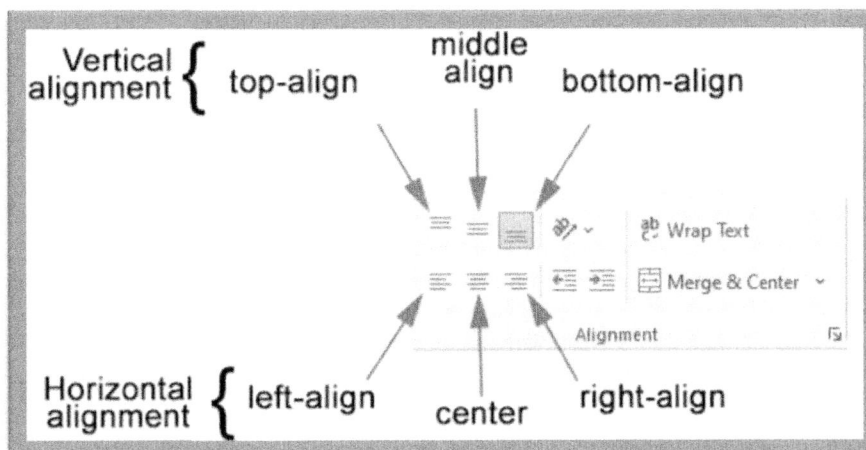

Alignment vertically. Choose one of the following icons to align data vertically:

- Top Align — aligns the cell's contents to the top.
- Middle Align — centers the cell's contents between the top and bottom.
- Bottom Align — aligns the contents to the bottom of the cell.

Alignment horizontally. For horizontally aligning your data, Microsoft Excel provides the following options:

- Align Left — aligns the cell's contents along the left edge.
- Center — places the items in the center of the cell.
- Align Right — aligns the cell's contents along the right edge.

Indent

In Microsoft Excel, the Tab key does not indent text in a cell as it does in Microsoft Word; it simply moves the cursor to the next cell. To change the indentation of the cell contents, use the Indent icons located just below the Orientation button.

To move the text to the right, click the Increase Indent button. If you've moved the text too far to the right, click the Decrease Indent button to move it back to the left.

Combine and Center

The cells merge style tool combines multiple cells in a worksheet into one.

The Merge and Center option is the quickest and easiest way to join two or more cells.

This is the simple two-step procedure:

1. Select the adjacent cells you want to merge.
2. Click the Merge & Center button on the *Home* tab > Alignment group.

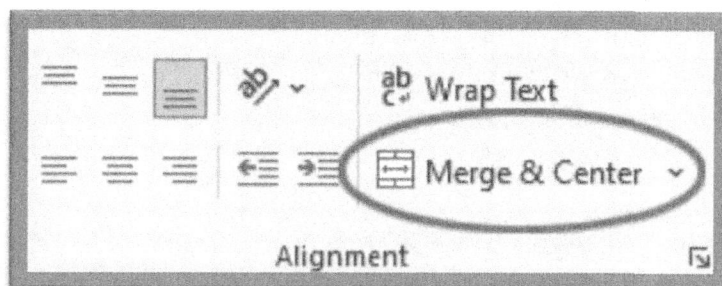

Applying Cell Styles

To format a cell or range with a different style:

1. Select the cell or range.
2. Select Home > *Styles*.
3. You can mouse over the different styles to get a preview on your worksheet before you select one.
4. Select a style from the pop-up menu.

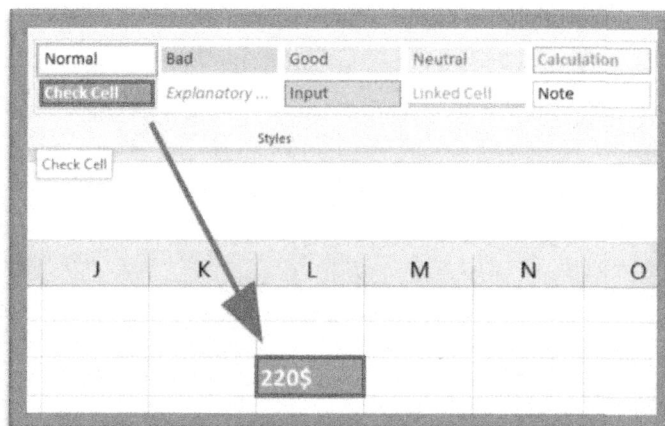

Chapter #6: The mathematics of Excel

It is critical to understand the distinction between a formula and a function in Excel.

A formula follows standard arithmetic syntax, with several functions, operators, and operands. The operators are the standard mathematical tokens (+; -; ; /...), and the operands are the formula data (1; 2; 3; 4;...). For instance, 1 + 5 + 3 + 4, or 2 3 5 10 and so on. The standard formula is manually entered into Excel using the operator.

The function, on the other hand, is preceded by a name and requires that you enter parameters. These are Excel functions, and openinusingg them is very simple. Simply insert =SU in the cell, for example, and you will get a screen with all the formulas available with the sum, or =RA and you open the functions beginning with RA. You can also go to Formulas and open the input window to select the desired function.

The Operators

Excel's operators are the basic arithmetic operators. They, like operands, control the terms of a formula.

The first thing to get used to when using Excel is to never lose sight of what appears in the formula bar indicated by the arrow in the image below.

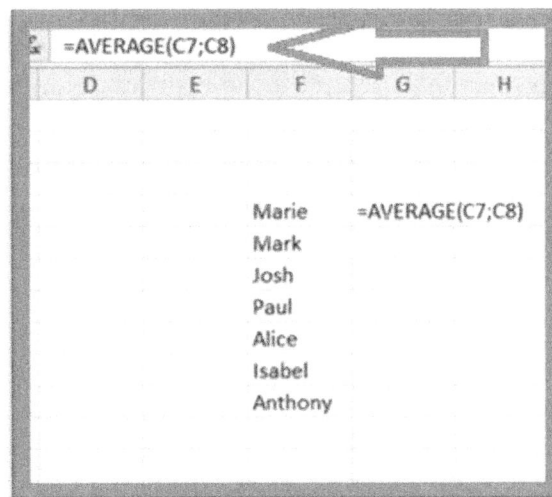

In that space you will see the operation you are performing, the formula you are using, and finally, the references, the selected cells, which will be the operands of the sum we are going to do.

The sum is straightforward: just put the term = in the cell of the last term you want to add, or at the bottom, if you want to add vertically, then select the first cell, enter the + sign, and continue with the items you want to add. This is an example to start learning the formulas, and before you know it you will see the many functions useful to you.

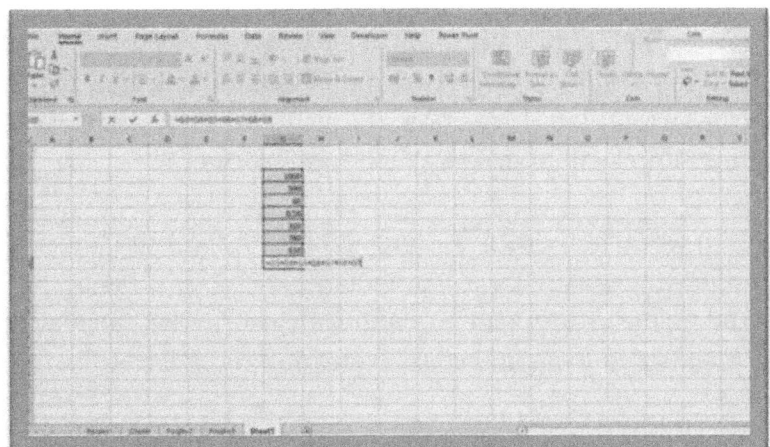

Unlike mathematical operators that perform an operation or put a formula into practice, logical operators give answers as true or false. Finally, the concatenate operator joins two strings; the concatenation operator is an ampersand (&).

Now let's see how to use it. The first crucial thing to remember is the necessary use of the formula bar, as seen in the images below. Let's say we want to merge the words "book" and "Excel". Our formula will be: ="book"&"Excel". As you can see below, Excel gives us a single word with no space.

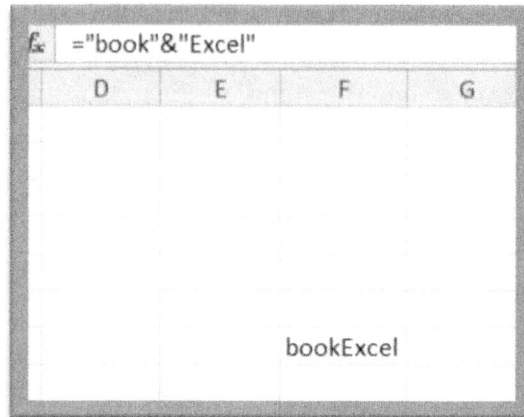

This premise behind the operators is critical for understanding Excel's priorities when reading a formula. Knowing the function of the operators in each category will help you understand the priorities and, as a result, the proper writing of a formula to achieve the desired result.

The BODMAS' primal logic is nothing more than mathematical expression logic:

- Brackets are important because they delineate the right of priority for operations in mathematical expressions.
- Then there's exponentiation, which happens in mathematical expressions as well.
- The priority in the four operations is division.
- Second multiplication
- Third addition
- Finally, subtract.

In Excel, what does this mean? Simply put, the software will determine the priority of the operations. Therefore, if you want a different priority, you must insert parentheses as you did in school. Remember when you had to start with an addition? You enclose them in square brackets. So it makes sense to do the thing with excel. Let's look at some guided exercises as examples now.

Assume we want to find the mean value between 7 and 13. We all know that the operation must be performed by dividing 7 + 13 by 2. To respect the BODMAS priorities, respect the mathematical priorities; thus, in Excel, enter the formula bar = (7 + 13) 2; the result is 10. If we remove the brackets, the result changes, and if we type = 7 + 132, Excel will return 13.5, which is incorrect.

Finally, if we do not remember the mathematical expressions well, we must always keep the BODMAS table in front of us when using Excel mathematics.

Experiment by inserting some mathematical formulas and learning to respect Excel's priorities, then write and solve the following mathematical expressions with Excel:

10 + 20 * 18 \s70-3 ^ 2

Average between 7000 and 3245.

Multiplication Table

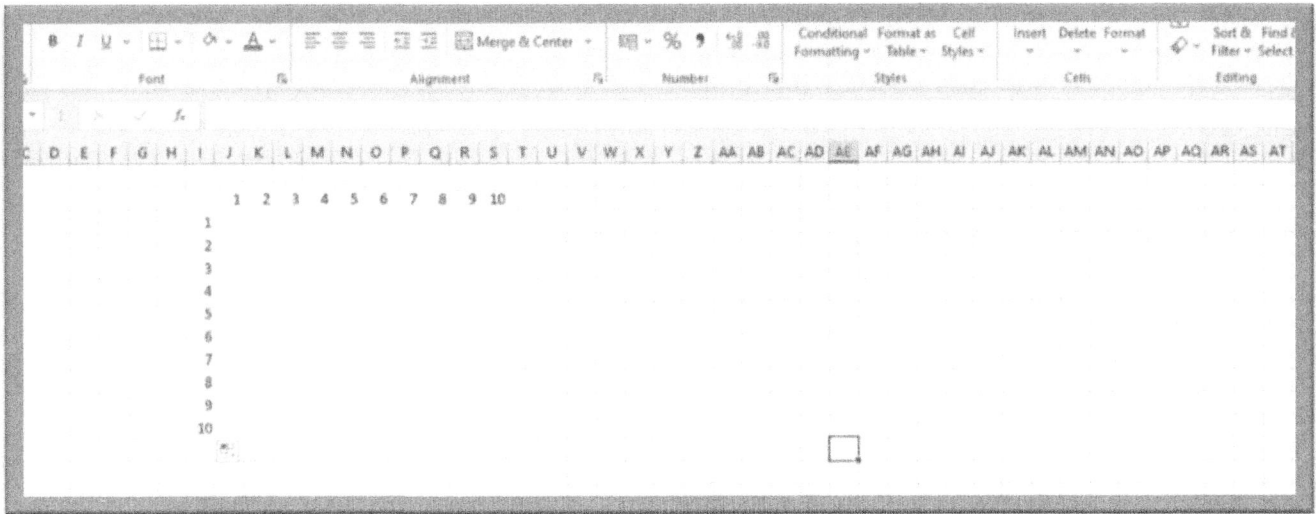

Let's start thinking about the first part of the formula: here we have to block the column. This means that it must always multiply the yellow column. To do this, we add the $ to the letter representing the column so we will write $ J2 * J2, but if J2 remains a relative reference, we will have the same problem. Therefore, in the second factor of this formula, we will block the row, adding the $ to the number that represents the row. We will find ourselves in front of this formula: $ I13 * J $ 2. We drag the first row horizontally, double-click, and have a perfect Multiplication table in two minutes.

Do not forget that if you use the F4 key, there will be very little room for error.

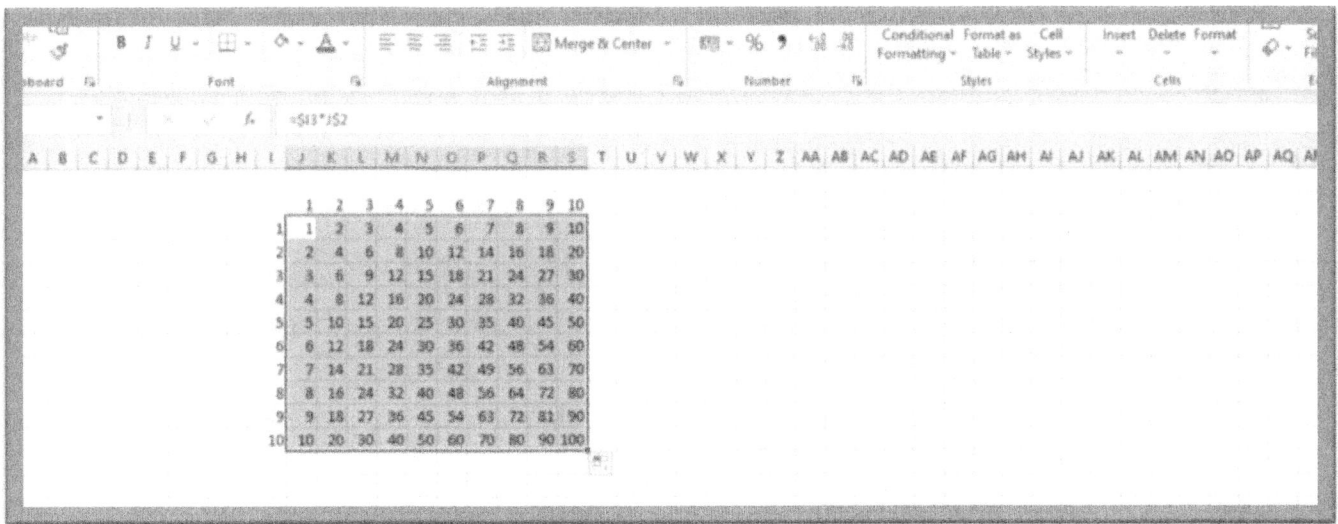

Learning references right away is very important for three fundamental reasons:

1. You can perform complex calculations quickly.
2. You are not bothered by complicated calculations.
3. Even though the F4 key directly inserts the formulas into you, as previously demonstrated, the margin of error is significantly low.

In Excel, do not get stressed or lose track, and orientation helps you work quickly and without making mistakes.

Average

This function is as easy as just getting an average of the numbers of the shareholders in the pool of a given company.

=AVERAGE (no.1, [no.2], ...)

Such as:

=AVERAGE (B4:B12) — this formula or function shows a straightforward average or (SUM (B4:B12)/10).

Subtraction

To execute the subtraction formula, in Excel, enter the cells to be subtracted in the format =SUM (B1, -C1). By putting a minus sign before that cell, one subtracts and can use the SUM formula. For example, if B1 is 12 and C1 is -5, =SUM (B1, -C1) returns 12 + -5, yielding 7. Subtraction, including percentages, is not a function or formula in Excel, but that doesn't mean it can't be done. Certain values can be deduced in two ways (or values inside the cells).

Division

To use the division formula in Excel, type =B1/C1 into the cells one wants to divide. This formula uses a forward cut to differentiate cell B1 by C1 by "." For example, if B1 is 5 and C1 is 10, the decimal value returned by =A1B1 is 0.5.

One of the most basic operations one will do in Excel is division. To do so, select an empty cell, type an equal's symbol, "=," and then the two values (or more) one wants to divide, separated by a forward dash, "." the outcome should be in the format =B2A2.

When one press enter, the quotient will display in the highlighted cell.

Percentage

To use the % formula in Excel, enter =B1/C1 into the cells for which a percentage is desired. To convert a decimal value to a percentage, highlight a cell, go to the home tab, and select "Percentage" from the dropdown menu.

While there is no specific Excel "Formula" for%, Excel makes it simple to convert the amount of a cell into a percentage, eliminating the need to estimate and enter the numbers.

Excel's home tab contains the basic settings for converting the value of a cell to a percentage. Select the tab, highlight the cell or cells to be converted to a%, and select conditional formatting from the dropdown menu next to it (the menu button may have previously said "General"). Then, from the resulting dropdown menu, select "Percentage." The presentation of each highlighted cell will be converted into a percentage. This element is located further down.

Please remember that if other formulas, such as the division formula (such as =A1/B1), are used to generate new values, the results will be displayed as decimals by design. Simply select the cells and change their type to "Percentage" from the tab before or after running the above algorithm.

Relative Referencing

The above image is the sample data set that we have. Let's say we have sales information for the three different types of apps for each quarter. The first thing we're asked to do is get the year's total for each app, so simply go to the total cell and use the shortcut key alt =. This automatically inserts the Sum formula (it's a handy trick as it realizes where the numbers are and it sums the values).

As the formula has picked up the values correctly, press Enter to calculate the total. Next, put your mouse on the game app total and use the fill handle to drag down; this will give you the answer for the other two columns. Let's look at our formula; if you click on the first row and click inside it, you'll see that it's summing B5 to E5.

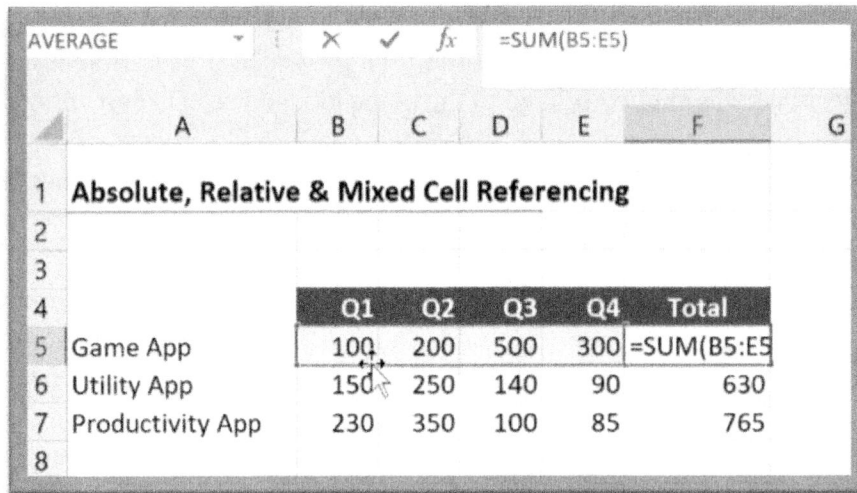

Now if you click on the last row and click inside the formula, you can see that it's summing B7 to E7.

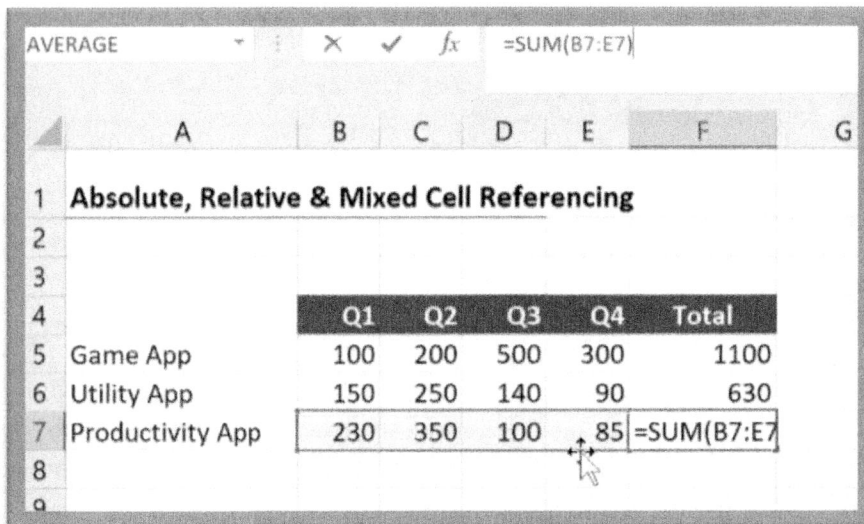

This is called relative referencing because the referenced cells are moving down as your formula moves down.

Absolute Referencing

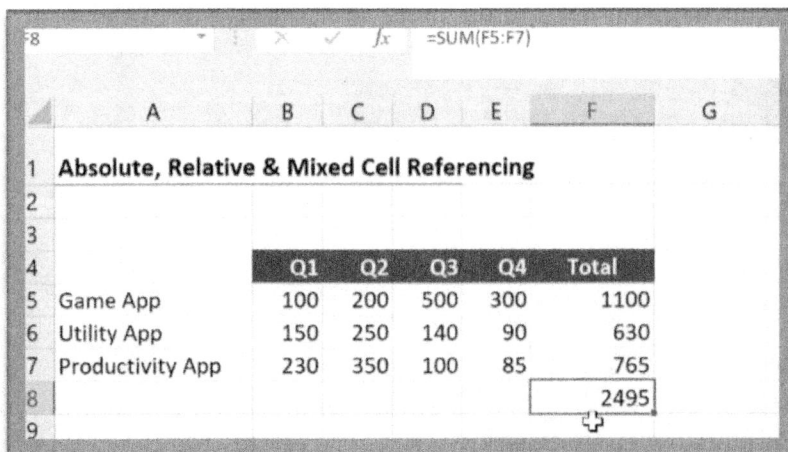

Assume your boss approaches you and asks, "Give me the percentage of each app compared to the year's total." The first thing you must do is compute the total. We already got the individual total in the previous reference, so we'll get the collective total by putting the mouse in the cell below the last total and using the same shortcut key combination alt + = and it'll automatically notice that it has numbers on top and sum them all. Simply press the Enter key.

Here comes the tricky part of calculating the percentage; to do this, go to the game app row and click on the empty cell next to its total, type in the equal sign, click on the first total for the game app, and divide it by the collective total.

We have to change the answer to a percentage. To do this, there is a search tab on the right side of the taskbar with "general" selected by default. Click the drop-down arrow and select percentage. Make sure that you click on the answer before clicking the percentage.

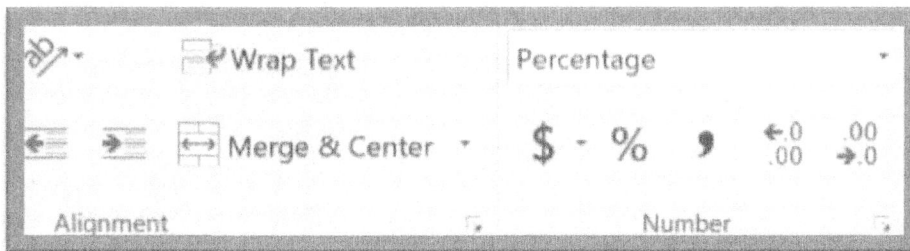

Now, if we pull down using the fill handle as we did before, we'll run into a problem.

Why do we run into this problem? Click inside the formula bar and you'll see that F6 is dividing F9

The total value decreased, which is correct because the percentage should reflect the utility app, but the total value should not decrease. We need to fix this, and this is when absolute referencing is required, which means adding dollar signs to both the row and column references.

Make sure you correct the first formula before pulling down.

Make sure you correct the first formula before pulling down

You can quickly add a dollar sign by pressing the shortcut key f4, which inserts the dollar sign into the formula. You can also manually enter the dollar signs if you don't want to use the shortcut key.

A quick tip: If you press F4 once, you will get the dollar sign for both the column and the row. If you click it again, the dollar sign appears only for the row. If you click it again, the dollar sign appears only for the column. After the fourth click, it returns to relative referencing. After putting the dollar sign, press ENTER, and when you pull it down, the formula will work properly.

The dollar sign fixes the column reference and the row reference; it doesn't move it down. When you click on one of the percentages, we can see that the first part is relative, and the second part of the formula is fixed to the collective total cell (F8).

Mixed Referencing

Assume your boss asks you to calculate the percentage of each quarter for each app so that we can quickly see which quarters have the most sales and which have the least sales for each app.

You will copy the initial information without the totals and percentages and paste it in a separate area but on the same worksheet.

You should now remove all the numbers because instead of these numbers, you want to see percentages, and the percentage is going to be the value you have in quarter one divided by the total you have in quarter one for that app.

First, input equal to sign in the new Q1 game app column and click on the previous Q1 game app column, this is to say that the new Q1 is equal to the old Q1. Divide it by the total of the Q1 row. When you do this, convert it to a percentage as we did in the previous example.

	Q1	Q2	Q3	Q4	Total				Q1	Q2	Q3	Q4
Game App	100	200	500	300	1100	44%	Game App		9%			
Utility App	150	250	140	90	630	25%	Utility App					
Productivity App	230	350	100	85	765	31%	Productivity App					
					2495							

For the first value, it's going to work fine, but when you pull it over to the right, the same thing that happened before when we calculated the totals occurs. Go to the formula bar, click on the f5, and press f4 to get it fully fixed. Now drag the Q1 over to the right, and it gives you the correct percentage for the other column.

	Q1	Q2	Q3	Q4
Game App	9%	18%	45%	27%
Utility App				
Productivity App				

Let's pull it down using the fill handle and see what it gives us.

	Q1	Q2	Q3	Q4
Game App	9%	18%	45%	27%
Utility App	14%	23%	13%	8%
Productivity App	21%	32%	9%	8%

Let's see if it's correct. As you can see, the utility and productivity app doesn't even add up to 100%, indicating something is wrong. Click inside the formula bar on the Q4 cell of the utility app to see that it's taking the Q4 number from the previous table, which is correct, but dividing it by the total for the game app. It's doing this because we fixed the reference to f5 completely, so it doesn't move at all, but we do want it to move partially, and the part we want to move is the row.

The dollar sign should not be on the part you want to move, but on the part, you want to be fixed. In this case, we want the F column to be fixed; it should never move to G or E. Instead, we want the row to move. This means that the dollar sign remains with the part we want to fix, so it remains with F, and you should remove the dollar sign from the formula bar row. We can now get it to work properly by dragging from Q1 to the end and pulling it down. You can test the accuracy by always going to the end of your data set and then double-checking that your formula referencing is correct.

Chapter #7: Functions

Here, we'll look at the top Excel formulae that every Excel user should be familiar with.

Mathematical Function

Math functions perform numerical activities such as percentages of totals, addition, and rudimentary financial analysis.

a)   The Sum Function

1.   The SUM function can add or sum values from multiple rows or columns.
2.   =SUM (num 1, [num 2])
3.   Follows these steps to use the SUM function.
4.   Create the SUM feature in the cell.
5.   To select cells for the cell range box, go to the Function argument.
6.   Then press the Enter key.

The SUM function takes the following inputs.

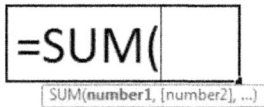

```
=SUM(
  SUM(number1, [number2], ...)
```

The function accepts the following arguments:

- **Number1:** This is the first numeric value to be added.

- **Number2:** This is the second numeric value to be added.

1.   Let's utilize the table and the Aggregate function to calculate the total income from Monday through Friday.

Days of Week	Sales
Monday	15000
Tuesday	21400
Wednesday	24541
Thursday	6521
Friday	15474
Saturday	13574
Sunday	15424

2.   Use Total to track income from Monday through Friday by following the steps below.
3.   To summarize, fill in the feature with the cell set in an empty cell; =SUM (A2:B6)

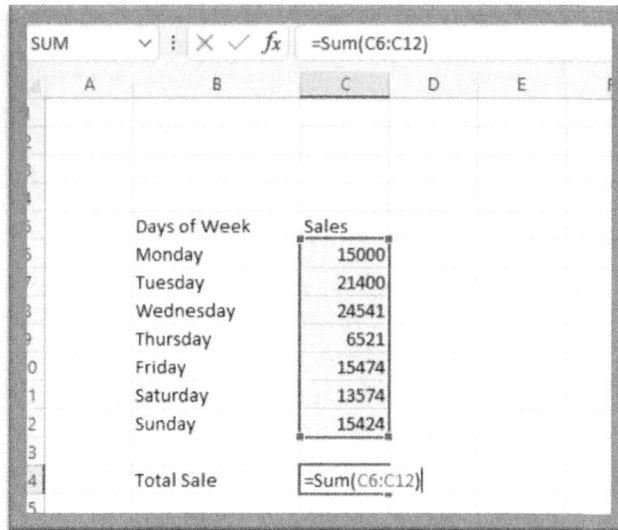

4. If you completed the parameters above, your net revenue from Monday through Friday would be 111934.

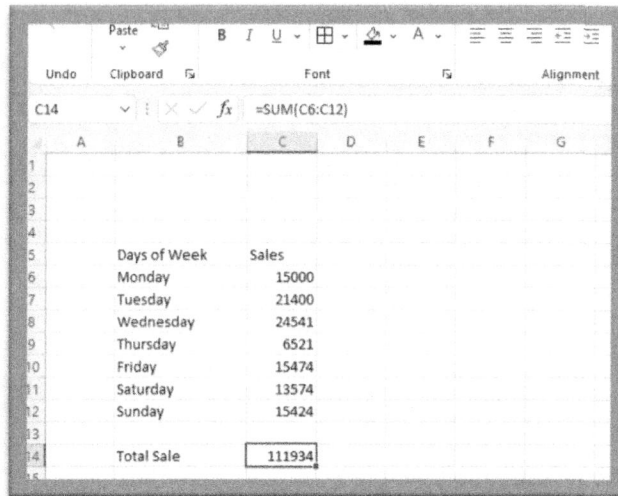

5. Take this into account when using the SUM function.
6. If the criteria supplied exceeds 255 characters, a value error occurs.
7. The SUM method immediately disqualifies empty cell types with text values.
8. You can use constants, sets, named ranges, and cell references as parameters.
9. The SUM function generates an error for each statement that contains errors.

b) Function SUMIF

The SUM function computes the sum of cells based on a set of parameters.

Dates, statistics, and sentences are used to create the criteria or requirements.

This work uses logical operators such as > and wildcards (*,).

The SUMIF method takes the following inputs.

=SUMIF (range, criteria, [sum_range]

The range of cells against which the criteria are expanded (Mandatory Argument).

Criteria (Required Argument): It determines which cells may be combined. Criteria arguments may be presented in several ways.

Numerical values include things like numbers, integers, and times.

Text strings include terms like Monday, East, Price, etc.

Expressions >11 and <3 are examples of expressions.

Sum_range (Optional Argument): This is the cell to sum if there are any further cells to sum than those specified in the range argument.

Let's examine whether the SUMIF function calculates revenue in January and the United States.

First, find out how much money you earned in January using the techniques listed below.

Fill in the role with an empty cell for the cell set to be rounded up. SUM (A2:A8)

Fill up the blanks for January's criteria. SUM (JAN, A2:A8,)

According to the data below, there were 81,037 sales in January.

To calculate the total number of sales in the United States.

With an empty cell, fill the role with the cell set to be summed =SUM (B2:B8).

The gross sales in the United States are shown in the table below.

Take this into account when using the SUMIF FUNCTION.

If the criteria supplied exceed 255 characters, a VALUE! error occurs.

Since the total range is not specified, the cells in range will be automatically summed.

If you don't use double quotes around text strings in parameters, it won't fit.

The SUMIF function's wildcards? There is also the option of using the symbol *.

c)   MOD Action

When a sum (dividend) is divided by another integer, the MOD function returns the remainder (divisor).

The MOD function is invoked using the following parameters:

That's the number you're looking for the remainder for. (Necessary assumption) (Mandatory Argument)

Divisor: That is the number by which you wish to divide the total.

Application of the MOD Function

Locate the remaining cell A2 in the table below using the MOD function.

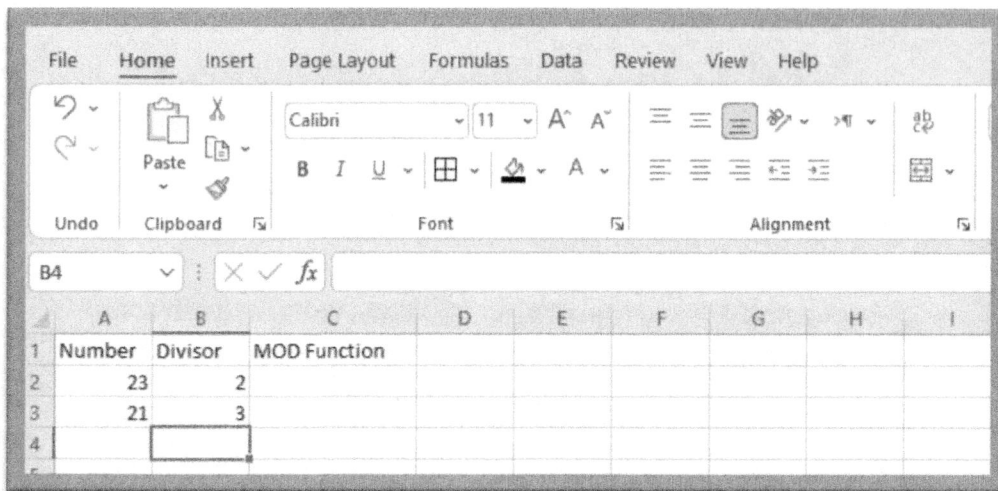

Locate A2 by following the steps below:

Write the function to be used, the integers, and the divisor =MOD (A2, B2) in an empty cell.

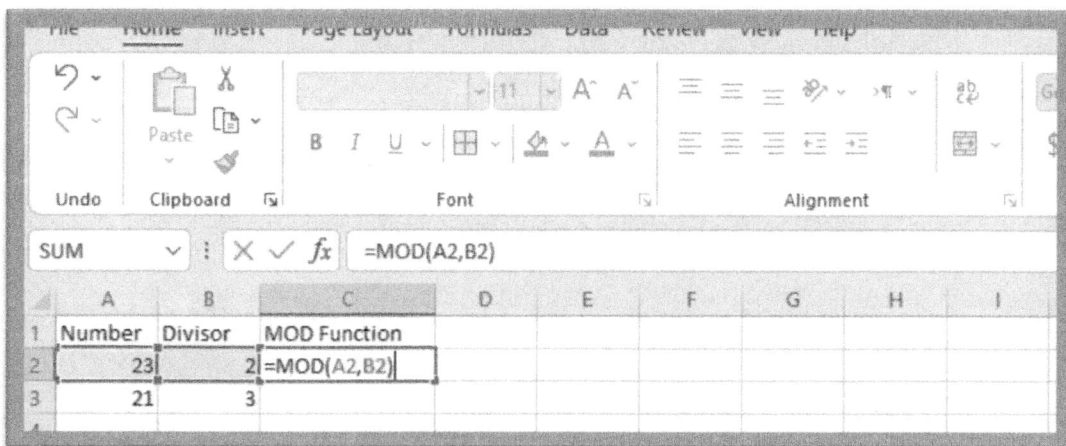

The consequence of the preceding move is shown in the diagram below.

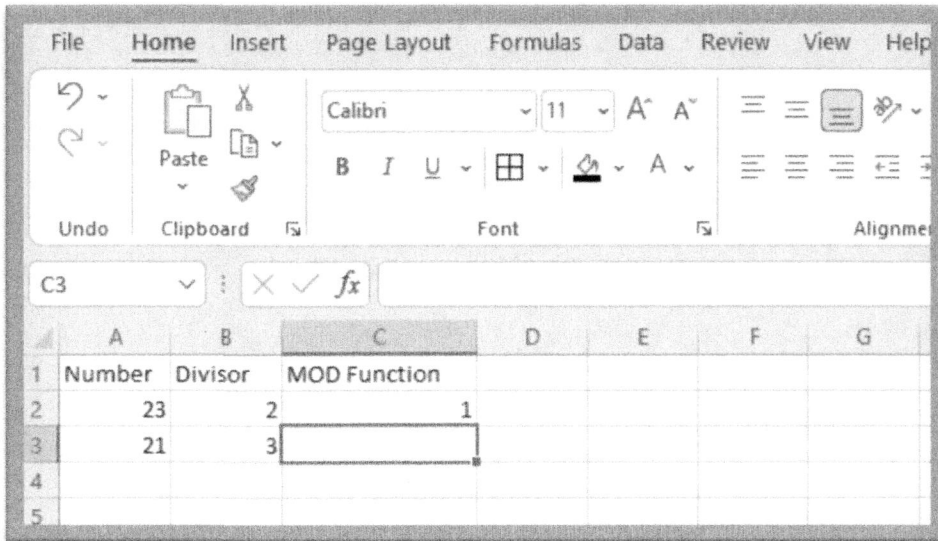

When using the MOD function, keep the following in mind:

#DID/0! If the divisor value is negative, an error occurs.

The result of the MOD function will have the same sign as the divisor.

d)   The Function RANDBETWEEN

The RANDBETWEEN function generates a random integer based on the input values. This functionality is activated whenever the worksheet is opened or modified.

The following arguments are passed to the RANDBETWEEN function:

Down: (Mandatory Function): The smallest number in the set that the function may return.

Peak (Mandatory Function): The highest integer in the set that the function can produce.

Use of the RANDBETWEEN Function

Let's look at how the RANDBETWEEN function is utilized in the table below.

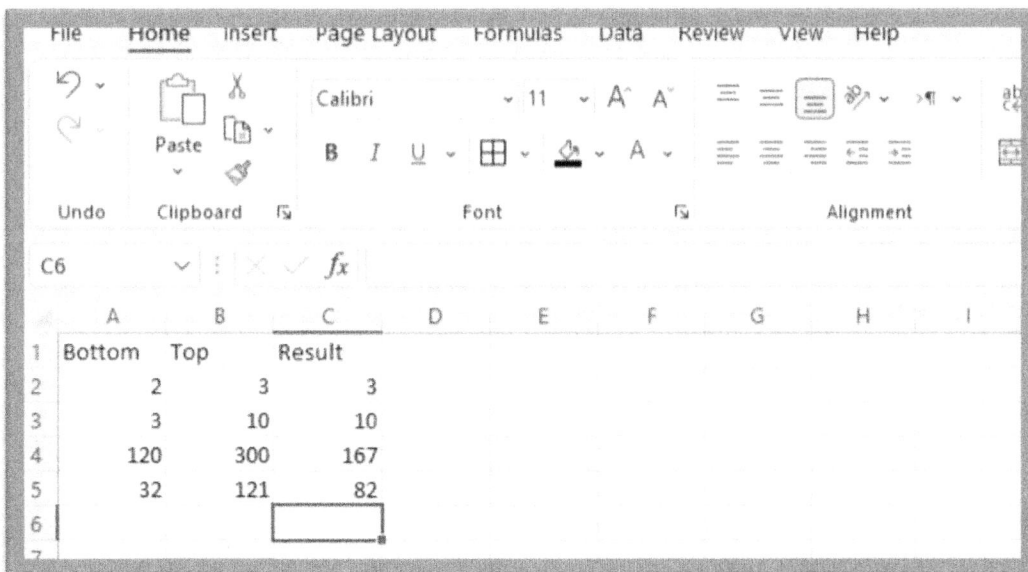

The table above includes the RANDBETWEEN technique. =BETWEEN (A2, B2).

The worksheet's result changes as the table's equations are repeated, as seen below.

Formula bar: =RANDBETWEEN(A2, B2)

	A	B	C	D
1	Bottom	Top	Result	
2	2	3	2	
3	3	10	8	
4	120	300	181	
5	32	121	87	

When utilizing the RANDBETWEEN tool, there are a few factors to keep in mind.

When the worksheet is tabulated or updated, the RANDBETWEEN function returns a new value.

Rather than shifting the random number as the worksheet is created, enter the RANDBETWEEN function in the formula bar and press F9 to convert the model to its output.

Choose a cell, enter the RANDBETWEEN module, and press Ctrl + Enter to generate a collection of random numbers in multiple cells.

e)   The Round Function

The ROUND function increases the number of digits in a number. You can use this function to round up or down. The previous parameters are used by the ROUND function.

Number 1 (Mandatory Argument): This is the number to round up to the nearest whole number.

Number of digits (Mandatory Argument): The number of digits used to round the figure.

Using the ROUND Function

Using the Round function, round 1844.123 to one decimal place, two decimal places, closest number, nearest 10, nearest 100, and nearest 1000.

To the nearest decimal place, invoke 1844.123 =ROUND (A1,1)

B1 formula: =ROUND(A1, 1)

	A	B	C	D	E
1	1884.12	1884.1			
2					

Round to the closest integer by entering 1844.123 =ROUND (A1, 0)

f)   ROUNDDOWN Function

The ROUNDDOWN function rounds values to a specified number of decimal places.

The previous inputs are used by the ROUNDDOWN function.

=ROUNDUP (Number, num digits)

Number 1 (Mandatory Argument): Round down to the nearest whole number with this figure.

(Required Argument) Several numbers: This is the number to round up to.

ROUNDDOWN APPLICATION

Using the ROUNDDOWN function, round 1233.345 to one decimal place, two decimal places, nearest number, closest 100, and closest 1000.

To round off, multiply 1233.345 by one decimal place. =ROUNDDOWN (A1, 1)

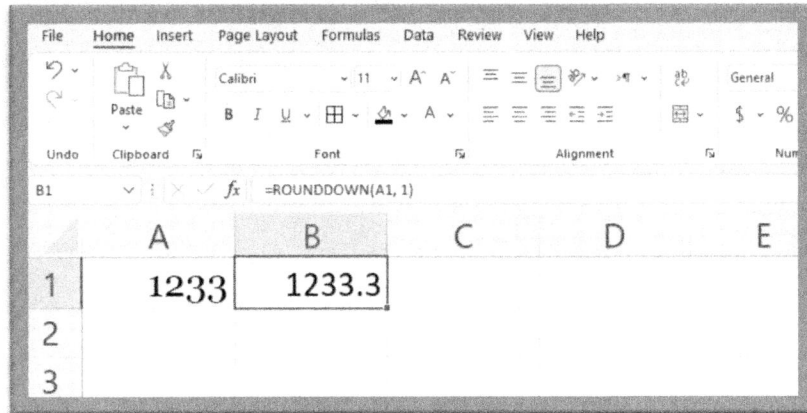

Round up to the next 1000th by 1233.345 =ROUNDUP (A1, -3)

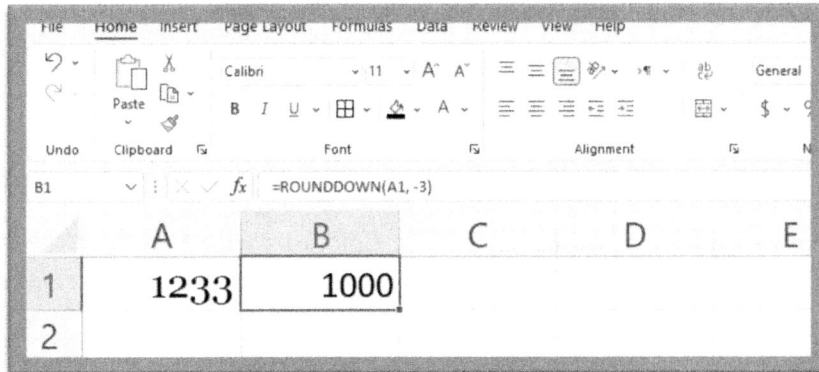

Counting and Analysis Functions

The counting and analysis functions used in Excel are:

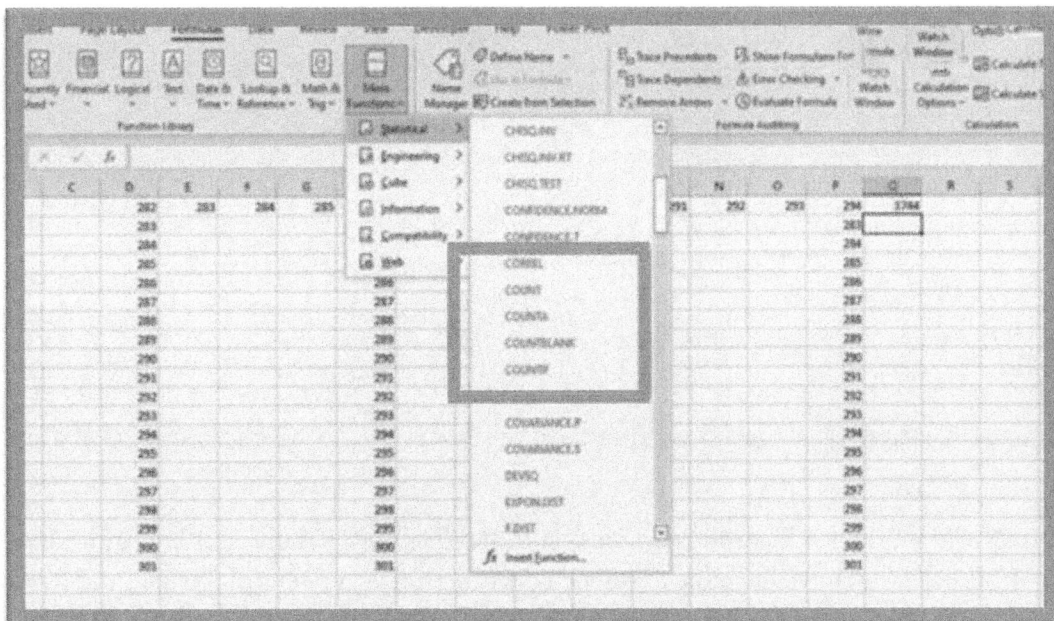

Count

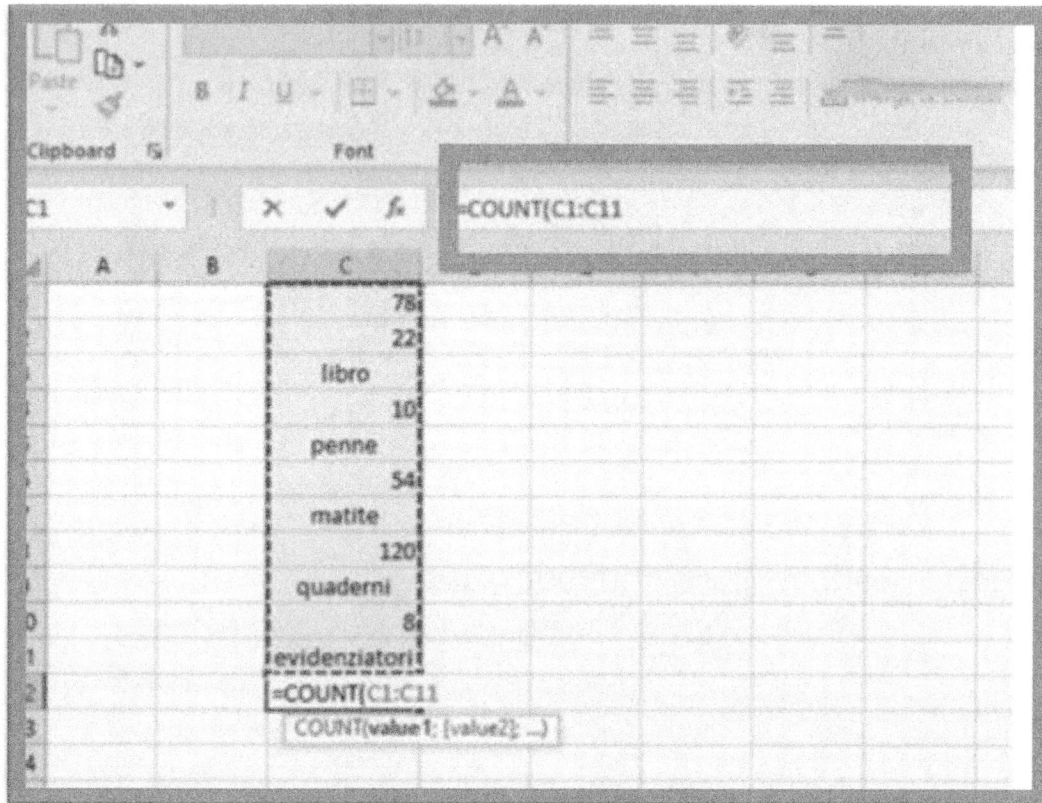

Well, let's start our journey into counting and statistical functions with COUNT. This function allows you to indeed count how many cells contain numbers. This can be useful when you need to extract data from a table.

Counta

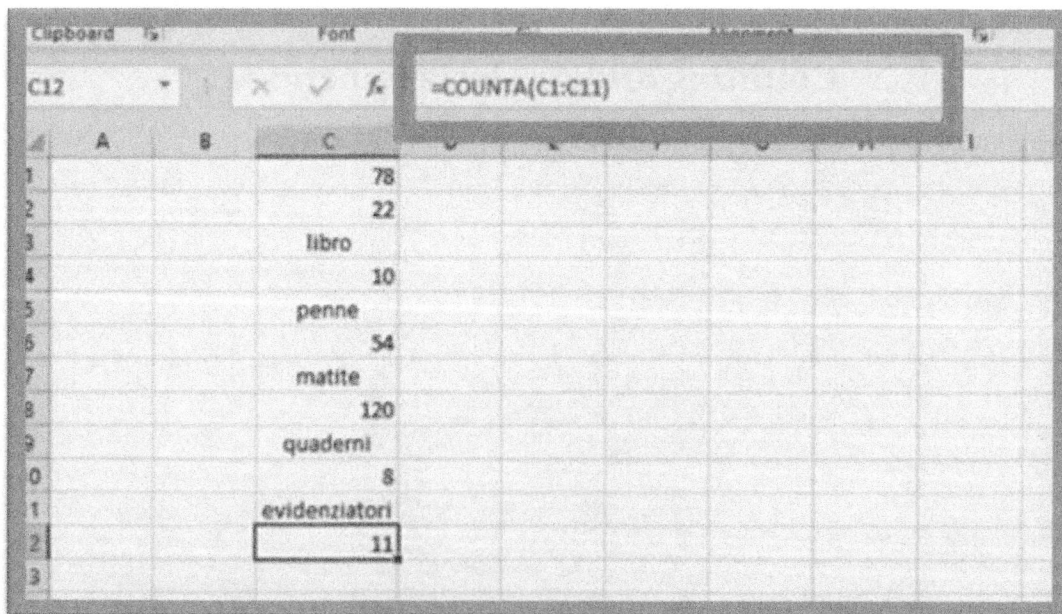

Countblank

The COUNTBLANK function produces the inverse result. It tells us how many cells have no values: empty cells. The function counts cells that have no value entered. This function is very simple to use, and the logic is similar to the previous two.

Countif

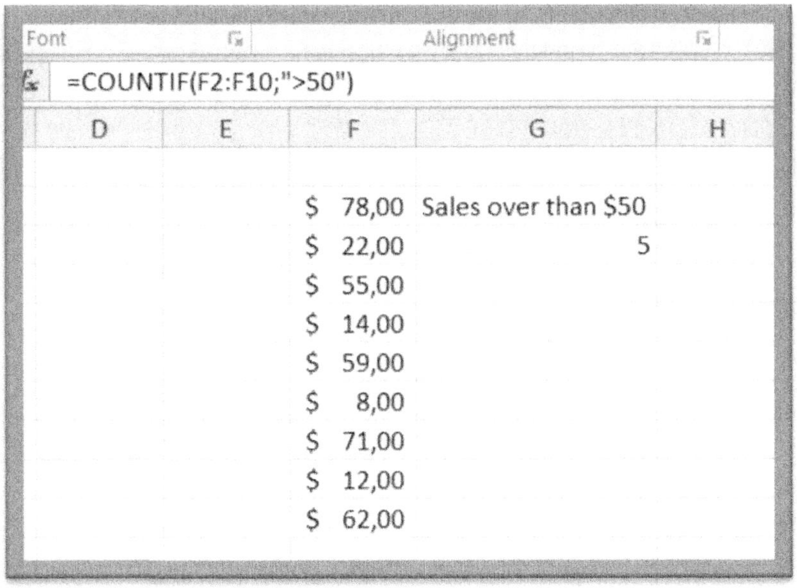

The COUNTIF function was introduced by Excel in its most recent versions. This function, not to be confused with the more well-known SUMIF function, is a statistical function that can be used to find specific data in a database. This function does not add up but it is still useful.

Countifs

The COUNTIFS function could be considered an evolution of the COUNTIF function as it adds criteria to it. While COUNTIF needs an interval and a criterion in its syntax, COUNTIFS is useful when we need to extrapolate data from a table using multiple criteria.

Date & Time Functions

Excel has a number of features for working with times and dates.
DATE, DAY, MONTH, AND YEAR TIME, HOUR, MINUTE, AND SECONDS DATED YEARFRAC AND IF
WORKDAY AND NETWORKDAYS EDATE AND EOMONTH
WEEKNUM AND WEEKDAY
Here's a quick rundown of the most important Excel functions to remember.
TODAY AND NOW TODAY and NOW functions can be used to get the current day and time.

Since the NOW function technically returns the current date and time, you can format it as a time only, as seen below:

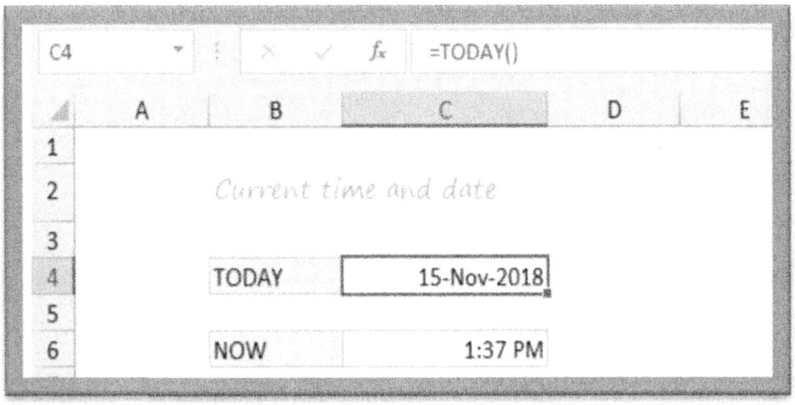

NOW () RETURNS CURRENT DATE

CURRENT DATE RETURNS TODAY ()

Please keep in mind that these are volatile functions, which means they will recalculate every time the worksheet is changed. If you want a constant value, use date and time shortcuts.

THE DATE, THE DAY, THE MONTH, AND THE YEAR

With the DAY, MONTH, and YEAR functions, you can disassemble any date into its raw components and then reassemble it with the DATE function.

=DATE (2018,11,14) / yields 14-November-2018 = DAY ("14 November 2018") / returns 14 =MONTH ("November 14, 2018") / returns 11 =YEAR ("November 14, 2018") / returns 2018

HOUR, MINUTE, AND SECOND

Excel includes several time-related parallel functions.

You can extract pieces of time using the HOUR, MINUTE, and SECOND functions, and you can generate a time from individual components using the TIME function.

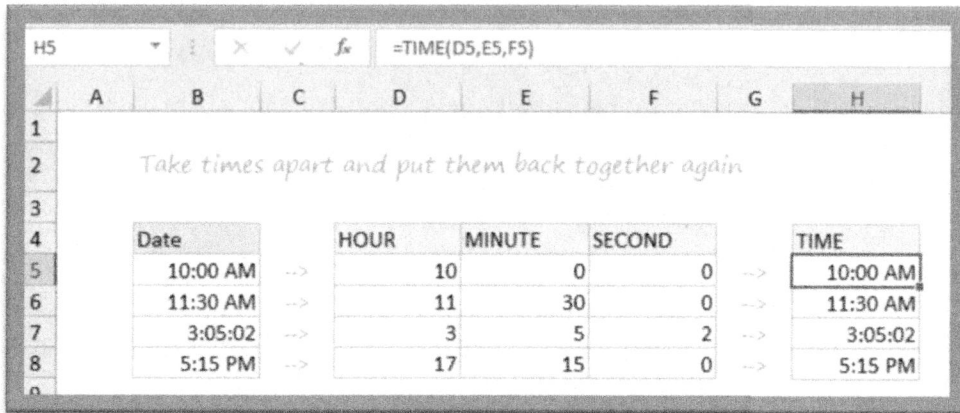

=*TIME* (10,30,0) // returns 10:30   =*MINUTE* ("10:30") // returns 30

=*HOUR* ("10:30") // returns 10   =*SECOND* ("10:30") // returns 0

WEEKDAY and WEEKNUM

The WEEKDAY function in Excel may be used to determine the day of the week from a date. WEEKDAY returns a number ranging from 1 to 7, indicating Sunday, Monday, Tuesday, and so forth. To find the week number in a given year, use the WEEKNUM function.

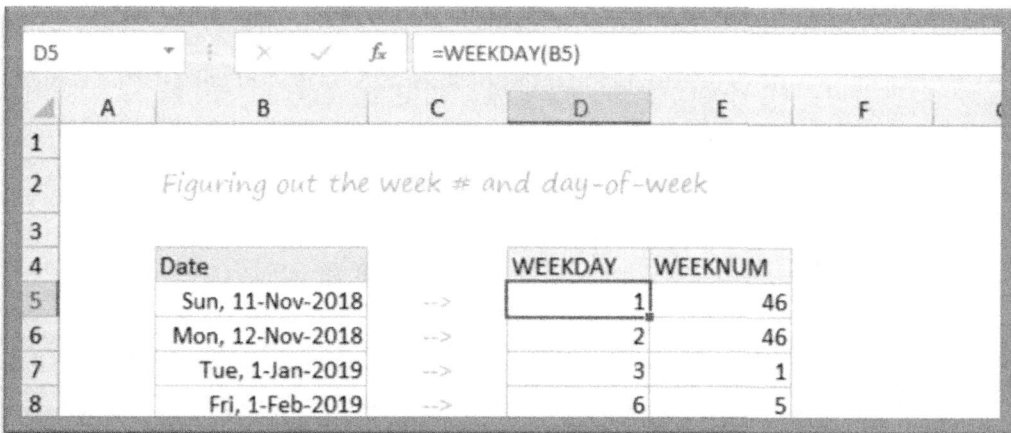

=*WEEKDAY* (date) // returns number 1 to 7

=*WEEKNUM* (date) // returns week number in year

Sorting Function

Sorting the output of a column in ascending or descending order is done using the Sort function. The SORT function makes use of the previous statements.

=SORT (array, [sort_index], [sort_order], [by col])

(This is a necessary argument.) This is the collection or sequence of values that will be filtered out.
(Supplementary Argument) Sort index: It specifies which column or row should be sorted.
(Supplementary Argument) Sort _order: This is the number used to order the cells; 1 means ascending, while -1 means descending. The results will be sorted in ascending order if this section is skipped.
(Supplementary Argument) by col: This controls the sorting orientation, with FALSE indicating row filtering and TRUE indicating column sorting.

Application of SORT Function

Using the SORT algorithm, we will sort the cells in ascending order in the table below.
Beginning with the lowest item and working your way up we will sort in ascending order using the steps below.
Type the function (=SORT), the source array (A2:B8), the sort of index (2), and sort order to see what happens (1). In conclusion, the formula =SORT (A2:B8, 2, 1) will be entered into an empty cell by clicking on it.
The data will be examined in ascending order after you hit Enter.

To arrange in descending sequence, from highest to lowest.

By clicking on an empty cell, you may insert the feature to be utilized. The sort of index (2), the root list (A2:B8), and the sort order are all equal to SORT (1). Last but not least, =SORT (A2:B8, 2, - 1).

The data will be sorted in ascending order when you click Enter.

	A	B	C	D	E	F
1	Item	Qty.		Pears	40	
2	Apples	30		Oranges	36	
3	Cherries	29		Lemons	34	
4	Grapes	31		Grapes	31	
5	Lemons	34		Apples	30	
6	Oranges	36		Cherries	29	
7	Peachs	25		Peachs	25	
8	Pears	40				

D1   fx   =SORT(A2:B8, 2, -1)

What You Should Know About the SORT Function

The SORT algorithm sorts items in ascending order by utilizing the first column as an example.

The SORT function is only available to Microsoft 365 members.

The output is automatically updated when the source data changes.

Chapter #8: Graphs

Excel Charts

There's a lot you can do with data, including making more than a single graph. By pressing a button to move the chart display from one display to another, you can compare data differently.

For example, the graph you build compares and contrasts salespeople. To illustrate how each salesman compares to the others, Excel divides data by spreadsheet columns and contrasts spreadsheet rows.

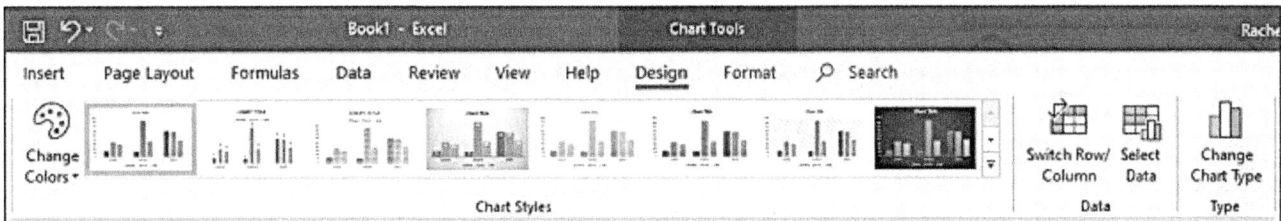

Creating a Chart

You can create either an embedded chart or a chart sheet.

a)   Select the cells containing the information you want to see in the chart. If you want the column and row labels to appear in the chart, make sure to include them in your selection.

b)   Choose Chart Wizard.

c)   Follow the chart wizard's instructions.

Make a Chart in a Single Step

a)   Select the data to be plotted, then press F11 to generate a chart sheet with the default chart type.

b)   Select the data to be plotted and then click Default Chart to generate an embedded chart with the default chart type. If it isn't already there, add the Default Chart button to a toolbar.

Make a chart out of nonadjacent choices.

a)   Select the first set of cells in the table that contain the data you want to enter.

b)   While holding CTRL, select any additional cell groupings you want to include. The nonadjacent selections must form a rectangle.

c)   Select Chart Wizard.

d)   Follow the Chart Wizard's instructions.

Add Chart Titles

It's a good idea to provide descriptive titles for the Chart so that viewers don't have to guess what it's about. Titles can be assigned to the chart and the chart axis, which quantify and explain the chart data. The graph contains two axes. The longitudinal axis runs through the left side (also called the value or y-axis). The numerical scale that can be used to represent column heights is represented by this axis. Right on this horizontal axis are the months (also called the category or x-axis). To quickly add chart details, select the Chart, then go to the Design tab and search for Simple Layout in the "Chart Layouts group". Each option displays a separate interface that changes how the chart elements are displayed. As shown in the diagram, Layout 9 includes stand-ins for a chart title or axis names.

The titles are entered directly into the Chart. The title of this Chart is the name of the drink, *Northwind Traders Tea*. The title of the y-axis on the left is Cases Sold.

First Quarter Sales is the title of the x-axis at the right. On the Design page, in the Chart Designs group, you can also insert names. By selecting Add Chart Component and Axis Names, you may add titles to your Chart.

Select a different chart type
You can alter the chart type of a data series or the entire chart in most 2-D charts. You can only modify the type of the entire chart with bubble charts. Changing the chart type impacts the entire chart in most 3-D graphics. A data series can be changed to a cylinder, cone, or pyramid chart type for 3-D column and bar charts.
a)   Choose one of the following options: Click the chart to change the chart type for the entire chart. To change the chart type of a data series, click the data series.
b)   Select Chart Type from the Chart menu.
c)   Select the chart type you want from the Standard Types or Custom Types tabs.
To apply the cylinder, cone, or pyramid chart type to a 3-D column and bar data series, choose the Apply to selection check box and then click Cone, Cylinder, or Pyramid in the Chart type box on the Standard Types tab.

Saving and Retrieval of Charts
It's similar to completing the operation of creating, saving, and retrieving data and charts all at once without exerting additional effort because the sheet will be recovered automatically while recovering it. When you invoke the graph after providing more efficient views, you can see that it creates a new sheet with the name chart, and changing the type increases the number of chart sheets. Other features available in Excel, such as pivot tables and online forms, should be discussed.

Selecting a Chart Style

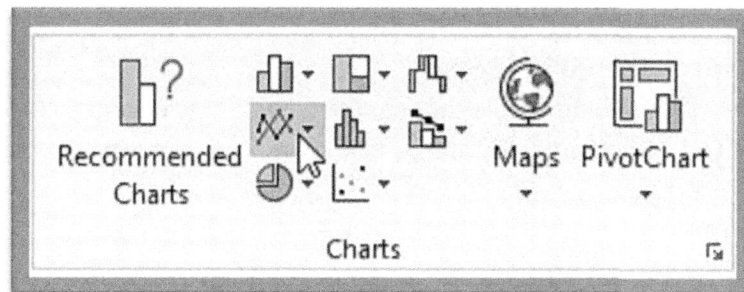

Different graphs highlight different details. You may alter the appearance of a chart after it has been developed. You may easily apply an already defined layout and design to your Chart instead of manually inserting or modifying chart elements or styling it.

A pie chart is used to compare values within a single collection and to show how different sections relate to the overall picture.
A line chart is the most straightforward way to depict patterns and changes over time. Use a Line chart if you want dates at the bottom of the chart to see past developments at a glance. A line map typically has only one collection of numbers on the y-axis.
An XY Scatter map compares two sets of figures, one on the x-axis and one on the y-axis, at the same time. The data are scattered throughout the graph. You can connect the attributes to the line, but those lines will not show you any patterns over time.

XY Scatter plots are useful for displaying numerical comparisons, such as those found in empirical or mathematical statistics, where multiple figures must be displayed on a single graph. For example, an XY Scatter map can be used to show the number of flu cases by age group or the average earnings of cities of various sizes.

Creating a Graph

Despite the fact that charts and graphs are two different things, Excel classifies all graphs into the chart types specified in the preceding sections. Follow the instructions below and pick the proper graph type to build a graph or another chart type.

Selection of Range

By moving your mouse over the cells containing the data you wish to utilize in your graph, you may highlight them.

The grayed-out cell area will now be illuminated.

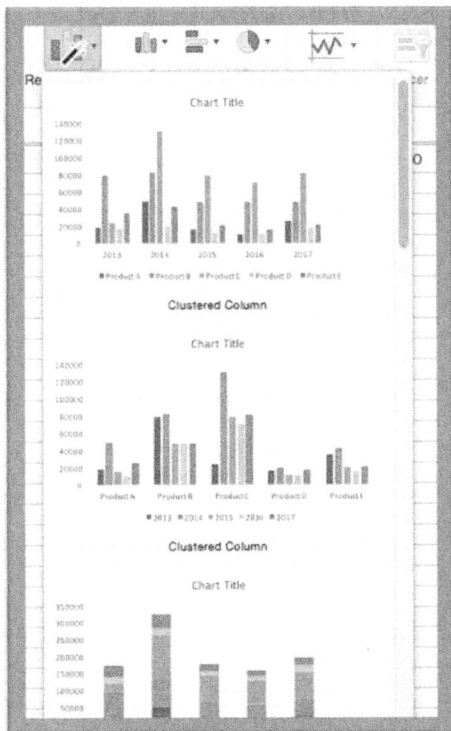

On the toolbar, choose Recommended Charts from the Insert tab. Then choose the graph type you want to use.

To personalize your graph, repeat the methods outlined in the preceding section. When constructing a graph, all of the functionality for producing a chart stays the same.

Adding the Sparkline

Sparklines allow you to see broad patterns in your data without having to create a full-fledged graph and chart. Sparklines are mini-graphs embedded in one cell that represent the cells in the section that follow it.

Return to the original data, navigate to the "Insert" page, and select a line, column, or loss/win sparkline by selecting the appropriate icon in the Sparklines chart.

Once you've selected the sparkline to include, a "Create Sparklines" dialogue box will appear, asking you to specify the "location Range" and "data Range." Make sure your mouse is flashing within the "Data Range" box, then click and drag on the Data you want to pick and unlock. The cells you selected earlier should now be filled in the "Data Range" window. Then there's the "Location Range," which allows you to specify where you want the Sparklines to appear. Check that the cursor in the Location is still flickering. Move your cursor over the cells in the range box, then release. A "Location Range" box would be automatically filled in. "OK" should be chosen.

Sparklines will appear in the cells you previously selected. Additionally, a new toolbar may appear.

Sparkline's Tools

There are numerous options available under the Sparkline Tools tab. You can change the type of Sparkline by going to the Form category and selecting a line, column, or win/loss.

To offer the Sparkline a bit more detail, you may even apply high or low points. To do so, go to the display category and choose one of the six "Sparkline Tools" choices.

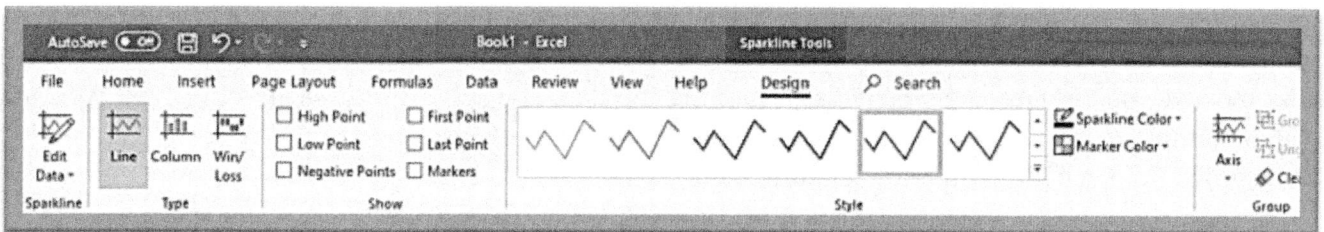

You may either adjust the hue by going to a "Sparkline Color" button mostly to the right and choosing a color or selecting the down arrow in the design group and choosing a suitable color. If you want to include low or high points, a different marker color will help them stand out. Go to the design group and press the marker color icon to adjust the marker color. When you choose a color, you'll see that the points on the panel will have become more distinct.

Move and Resize an Embedded Chart

An embedded chart is a movable, resizable, and copyable representation. The selection controls will show when you click the embedded map to render it operational.

To zoom into the Chart, follow these steps:

• Retain the Shift key when dragging one of the range handles to maintain the map proportions as you resize.

• Render the map active and then drag the cursor over a blank field to move it. Release the button of the mouse after dragging the embedded map to the new spot.

Resizing and Moving Tips

Ensure you have chosen the whole Chart, not only one of the chart parts, to activate it. The collection handles, which will sit at the Chart's outermost margins, will indicate this.

Once you've picked a blank region of the table, the mark chart Area may show as you pass the pointer over it. These pointers can assist you in selecting and moving the whole Chart, not just a single feature.

Shapes

To draw anything from shapes in Excel, choose any of the shapes you want to create, hold left-click drag, create your shape in the desired size, and then release the key to finish the drawing.

Smart Art

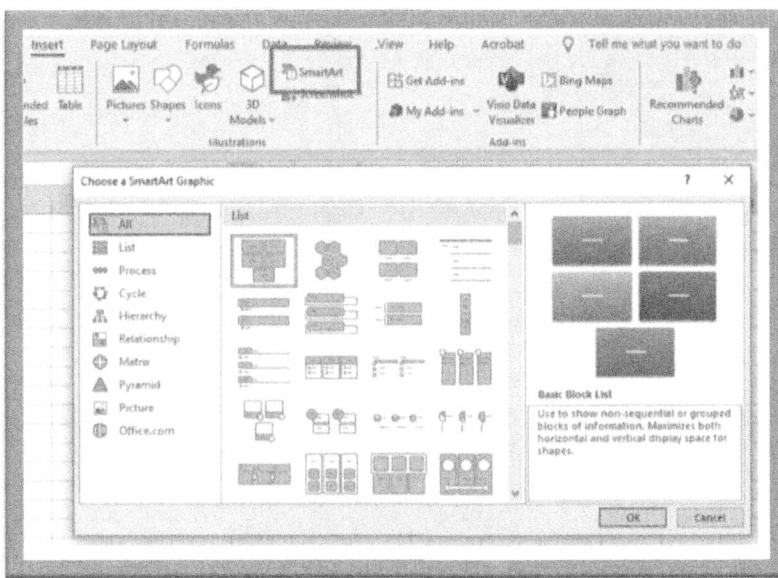

Smart Art Graphics are pre-made graphics in a spreadsheet or workbook displaying links, cycles, graphs, pyramids, and lists. These graphics do not contain or use pre-entered data from spreadsheets. Go to the Insert tab and select the Smart Art feature to add a Smart Art Graphic to your text (and type). The tab will appear in a highlighted green on your top toolbar, along with all of the graphic options, when you open it.

By pressing the button, you'll be able to insert a graphic into the spreadsheet you're working on. When you click on the image, a small dialogue box will appear, giving you the option to change the data that appears inside it. If you do not enter this dialogue box, the graphic will display the default text. If you close the dialogue box by accident, click the button on the left side of a graphic to bring it back up on the screen.

Adding a Smartart

Choose a design that appeals to you.

The Smart art tool allows you to look up the definition of any word in any cell of a workbook and conduct additional research on it. You can still search for images within the Excel workbook using this command. If you want to use Smart art, click the cell with the word, then the Review tab. The next step is to select the Smart art command. The right margin of your workbook will display information about the word.

There are Define and Explore headings available. If you select Explore, it will give you research on the word, including images. The source of information is quoted.

Images

In your Excel spreadsheet, click where you want to insert a picture.
Switch to the Insert tab > Illustrations group and click Pictures.

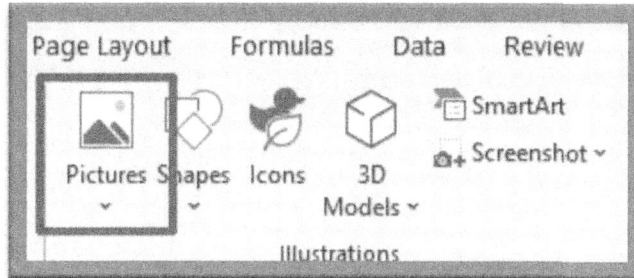

In the Insert Picture dialog that opens, browse to the picture of interest, select it, and click Insert.

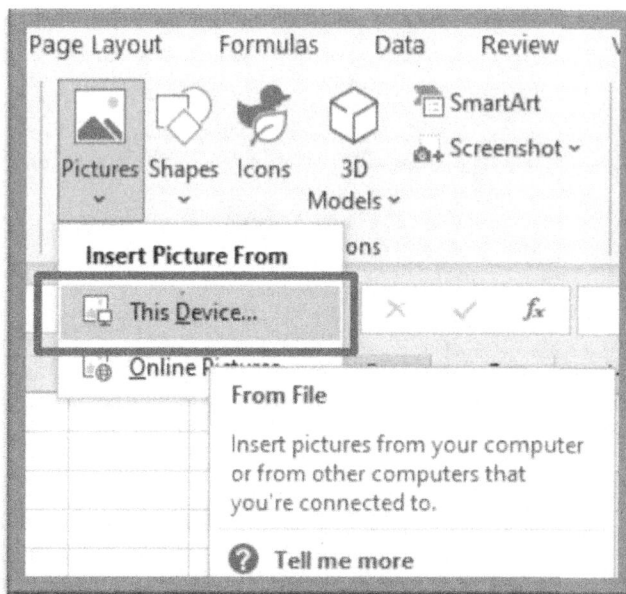

Add Artistic Effects to Images

Like the most popular photo editing software programs, you can easily add effects to your images with Excel.

Double-click your image to open the Format tab, then choose the Artistic Effects option.

Select the creative effect you wish to use in your photograph.

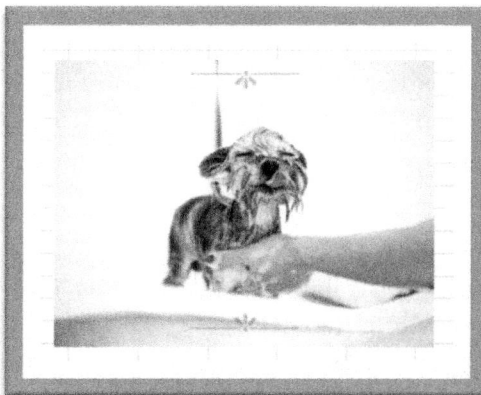

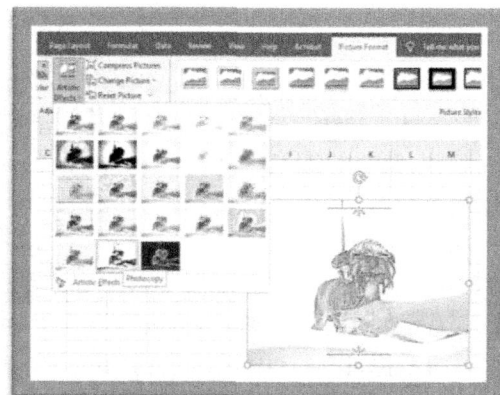

PART TWO: EXCEL FOR INTERMEDIATES

Chapter #9: Filters

You can use the FILTER function to narrow a range of data based on the criteria you specify. This will produce an array of values that match the criteria. The filter searches data using logical tests to find records that meet a set of predefined criteria. A FILTER can be used to match data within a specific time frame, to contain specific text, or to be a function of a specific value.

FILTER takes three arguments: array, include, and if empty. A range to filter is represented by an array. The argument should consist of one or more logical tests. Based on the value evaluated from the array, these tests should return TRUE or FALSE. "If empty" specifies the result to be returned when no matches are found by FILTER. This is usually a message like "No records found," but other values can also be returned. If you don't want to see anything, enter an empty string ("").

Using FILTER, you will get a variety of results. The FILTER algorithm will automatically update the results when the source data value or the size of the source data array changes. The results of FILTER will appear in multiple cells on the worksheet.

Filter

Filter the Excel files if you only want to view papers that match specified requirements.

1.   Select each cell in a data collection by clicking on it.

2.   On this Data screen, choose Filter from the Sort and Filter group.

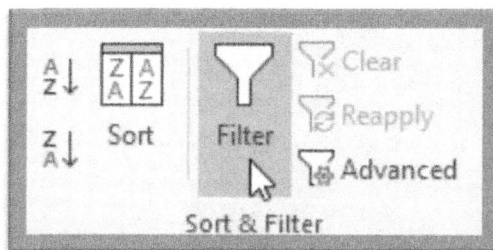

3.   There are arrows in the column headings.

	A	B	C	D	E
1	Last Nan ▼	Sales ▼	Count ▼	Quart ▼	
2	Smith	$16,753.00	UK	Qtr 3	
3	Johnson	$14,808.00	USA	Qtr 4	
4	Williams	$10,644.00	UK	Qtr 2	
5	Jones	$1,390.00	USA	Qtr 3	
6	Brown	$4,865.00	USA	Qtr 4	
7	Williams	$12,438.00	UK	Qtr 1	
8	Johnson	$9,339.00	UK	Qtr 2	
9	Smith	$18,919.00	USA	Qtr 3	
10	Jones	$9,213.00	USA	Qtr 4	
11	Jones	$7,433.00	UK	Qtr 1	
12	Brown	$3,255.00	USA	Qtr 2	
13	Williams	$14,867.00	USA	Qtr 3	
14	Williams	$19,302.00	UK	Qtr 4	
15	Smith	$9,698.00	USA	Qtr 1	
16					

4.   Click the arrow next to a country to choose it.
5.   Press Select All and select a check box beside the USA to delete all checkboxes.

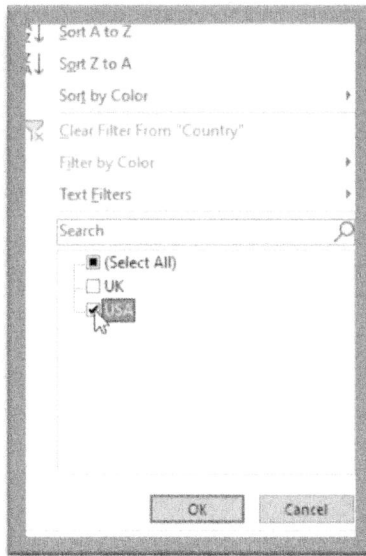

6.   Click the OK button.
7.   As a consequence, Excel only displays income from the US.

8.   Click the arrow next to Quarter to choose it.
9.   Press Select All and click the check box underneath "Qtr 4" to delete all checkboxes.

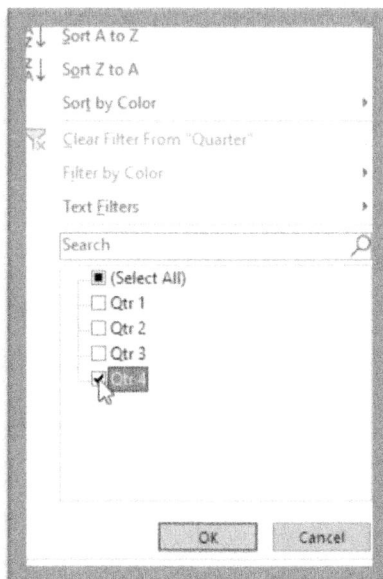

10. Select OK from the dropdown menu.

11. Consequently, Excel only displays income in the United States for "Qtr. 4."

	A	B	C	D	E
1	Last Nan ▾	Sales ▾	Count ▾	Quart ▾	
3	Johnson	$14,808.00	USA	Qtr 4	
6	Brown	$4,865.00	USA	Qtr 4	
10	Jones	$9,213.00	USA	Qtr 4	

12. Press Clear on the Data page to remove the filter from the Sort & Filter group. To remove the filter and the arrows, choose Filter.

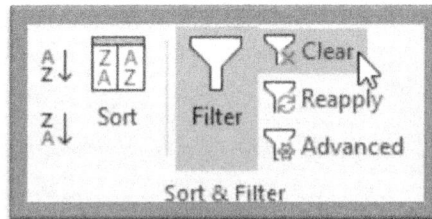

A↓Z Z↑A	Z↓A↑Z		▽x Clear
Z↑A	Sort	Filter	▽ Reapply
			▽ Advanced

Sort & Filter

13. Excel information should be filtered more rapidly.

14. Pick one of the cells.

	A	B	C	D	E
1	Last Name	Sales	Country	Quarter	
2	Smith	$16,753.00	UK	Qtr 3	
3	Johnson	$14,808.00	USA	Qtr 4	
4	Williams	$10,644.00	UK	Qtr 2	
5	Jones	$1,390.00	USA	Qtr 3	
6	Brown	$4,865.00	USA	Qtr 4	
7	Williams	$12,438.00	UK	Qtr 1	
8	Johnson	$9,339.00	UK	Qtr 2	
9	Smith	$18,919.00	USA	Qtr 3	
10	Jones	$9,213.00	USA	Qtr 4	
11	Jones	$7,433.00	UK	Qtr 1	
12	Brown	$3,255.00	USA	Qtr 2	
13	Williams	$14,867.00	USA	Qtr 3	
14	Williams	$19,302.00	UK	Qtr 4	
15	Smith	$9,698.00	USA	Qtr 1	
16					

15. Right-click and choose Filter, Filter using Value of Selected Cell from the context menu.

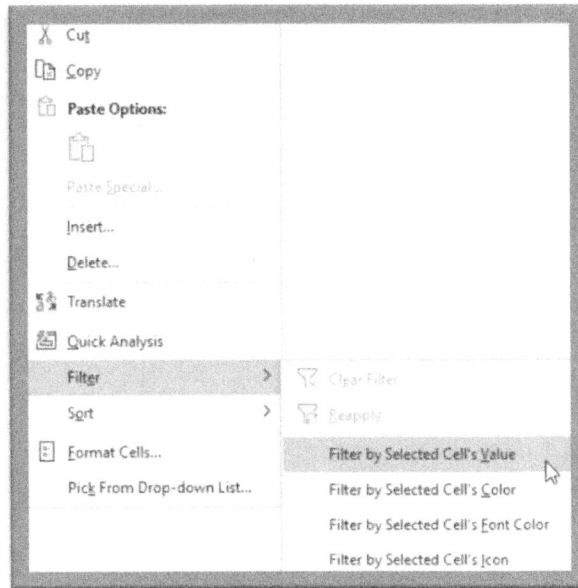

16.   As a consequence, Excel only displays income from the US.

	A	B	C	D	E
1	Last Nan ▾	Sales ▾	Count ▾	Quart ▾	
3	Johnson	$14,808.00	USA	Qtr 4	
5	Jones	$1,390.00	USA	Qtr 3	
6	Brown	$4,865.00	USA	Qtr 4	
9	Smith	$18,919.00	USA	Qtr 3	
10	Jones	$9,213.00	USA	Qtr 4	
12	Brown	$3,255.00	USA	Qtr 2	
13	Williams	$14,867.00	USA	Qtr 3	
15	Smith	$9,698.00	USA	Qtr 1	
16					

17.   Choose another cell from a different column to sort this data set further.

Sorting Data

Data sorting is an important part of data analysis. Arranging a list of names alphabetically, compiling a list of inventory levels from highest to lowest, or organizing rows by color or icon is very useful. Organizing and locating the appropriate data can assist you in quickly visualizing and understanding data, making better decisions, and ultimately making sense of your data.

Data can be sorted using alphabetical order (from A to Z or Z to A), numerical order (from smallest to largest or largest to smallest), and chronological order (from oldest to newest or newest to oldest). A custom list can be sorted by size (for example, large, medium, or small) or format (for example, font color, cell color, or icon set).

1.   The text should be sorted.

2.   Select a cell to sort the column.

3.   Click Sort A to Z to quickly sort in ascending order.

4.   Sort quickly from Z to A by clicking "Sort Z to A."

Potential Issues

Check that all data is saved as text. In a column that contains both numbers and text, you must format either the numbers or the text. If this format is not used, the numbers stored as numbers are sorted before the text numbers. In order to format selected data as text, open the Format Cells dialog, press Ctrl+1, then select the Number tab, then General, Number, or Text under Category.

There should be no leading spaces. Leading spaces are sometimes inserted before data is imported from another application. Remove any leading spaces before sorting the data. You can use the TRIM function to do this, or you can do it manually.

1. Numbers should be sorted.
2. Select a cell to sort the column.
3. Go to the Data tab and perform one of the following actions in the Sort & Filter group.
4. By selecting "Sort From Smallest to Largest," you can sort from lowest to highest.
5. By selecting "Sort by largest to smallest," you can divide larger items into smaller ones.

Sorted Dates and Times

1. Select a cell to sort the column.
2. Go to the Data tab and perform one of the following actions in the Sort & Filter group.
3. To arrange items from an earlier to a later date or time, click on "Sort Oldest to Newest".
4. To arrange items from a later date or time to an earlier date or time, click on "Sort Newest to Oldest".

Note: If you are not obtaining the expected outcomes, verify that the dates and times are recorded as actual dates and times, as otherwise the column could have text. In order for Excel to accurately sort dates and times, they need to be formatted as a serial number of date or time. Dates and times that Excel is unable to identify as such will be saved as text.

Format the cells by weekday for sorting by days of the week. When sorting by weekday regardless of date, use the TEXT function to convert the cells to text. Because TEXT returns text values, the sorting operation will rely on alphanumeric data.

Sort multiple columns or rows at once.

When you have data that you want to group by the same value in one row or column before sorting the next row or column in the group of equal values, you may need to sort more than one column or row. Sort by department first, and then by name. This way, all employees in the same department are grouped together, and the names are alphabetized within each department. Sorting is possible across up to 64 columns.

1. Choose a cell from the data range.
2. On the Data tab, select Sort from the Sort & Filter group.
3. In the Sort by section of the Sort dialog box, select the first column that you want to sort.
4. Sort On allows you to choose the sort type. You can do one of two things:
5. By selecting Values, you can sort by text, number, or date and time.
6. Select Cell Color, Font Color, or Cell Icon to sort by format.
7. Select how you want the list to be sorted under Order. You can choose between the following options:
8. When selecting text values, select from A to Z or Z to A.
9. Choose Smallest to Largest or Largest to Smallest when comparing numbers.
10. Choose "Oldest to Newest" or "Newest to Oldest" for date or time values.
11. Choose Custom Lists to sort by the list you created.
12. Repeat steps three through five by clicking Add Level to add another column to sort by.
13. To copy a column to sort by, select it and then click Copy Level.
14. By selecting the column to sort by and then clicking Delete Level, you can delete it.

You can change the order of the columns by selecting an entry and clicking the Up or Down arrow next to the Options button.

The higher entry in the list is sorted first, followed by the lower entry.

Sort the results by cell color, font color, or icon.

If the colors in a table column or range of cells have been manually or conditionally formatted, they can also be sorted by the color of the cell or the color of the font. Sorting can also be done with a conditional formatting icon set.

1. Select the cell in the column that you want to sort.

2. Select Sort from the Sort & Filter group on the Data tab.

3. In the Sort Dialog Box, select the column to sort in the Sort by field under Column.

4. Select Cell Color, Font Color, or Cell Icon after clicking Sort On.

5. Click the arrow next to the button in the Order section, then choose a cell color, font color, or cell icon depending on the format you want.

6. Data can be sorted using text (A to Z or Z to A), numbers (smallest to largest or largest to smallest), and dates (oldest to newest and newest to oldest). If you have a custom list, you can sort it by size (e.g., large, medium, or small) or format (e.g., font color, cell color, or icon set).

7. Text should be sorted.

8. Select a cell to sort the column.

9. Select one of the following options from the Sort & Filter group on the Data tab:

10. Click Sort A to Z to quickly sort in ascending order.

11. Sort quickly from Z to A by clicking "Sort Z to A."

Check that all the data is saved with text. In a column that contains both numbers and text, you must format either the numbers or the text. If this format is not used, the numbers stored as numbers are sorted before the text numbers. To format selected data as text, press Ctrl+1 to open the Format Cells dialog, then click the Number tab and select General, Number, or Text from the Category drop-down menu.

Any leading spaces should be removed. Sometimes, data imported from another application will have leading spaces inserted before the data. Get rid of the leading spaces before sorting the data. The *TRIM function* can be used to do this, or you can do it manually.

Chapter #10: Intermediate level functions

The Lookup function allows you to find specific data in a table that you are looking for. This function was very popular in the past, but you have to consider its great limitation: it needs a list sorted in alphabetical order. It is now outdated by vertical VLookup and HLookup functions that Excel has released in other versions since 2003. When the 2007 version introduced the vertical search, which releases more precise results without having to meet specific conditions, the Search function has gone into disuse. The latest addition to Excel, the recent XLookup, is a really precise and simple tool to apply. This function will surely replace all the other Lookup functions over time, although it is not yet used on a large scale because it was recently released, with the new 365 suite.

However to give a more complete overview, it was necessary to mention this function. You can find it in Functions like all the others.

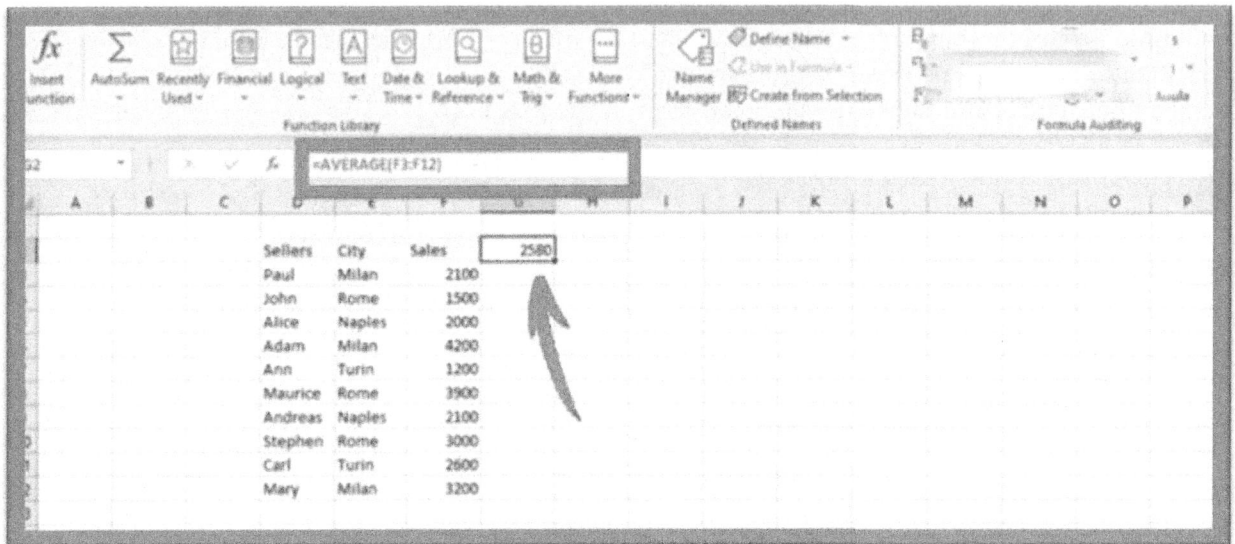

VLookup

VLOOKUP is one of the most useful and commonly used functions in Excel. It makes life much easier when working with very large databases, as without it, we would have to manually extrapolate data, wasting a lot of time.

As the name implies, we can use this function to start with a name and vertically recall what we need to know. The VLOOKUP, like all the others, has a syntax that must be followed for the program to recognize it and return the correct result.

The VLOOKUP syntax is as follows: Value; Table Array; col index num; [range lookup].

Here are the rules to follow:

1.   Since the selection must always go from left to right, the name column we give value to must be the first on the left. The name usually indicates the value.

2.   The names column, and the values column, must be unique. We must ensure that there are no duplicate names or Excel will generate an error message.

3.   We select the entire table, except the header row, in the table matrix to avoid errors.

Let's do a practical exercise now to see how we can do it together.

Let's create a database with the sellers' names and the number of closed sales contracts divided by month. Assume we now need to know how many contracts Paul, the sales agent, closed in September. This information can be quickly extrapolated from our database using the VLOOKUP function.

In the image, you will see our function highlighted in the formula bar.

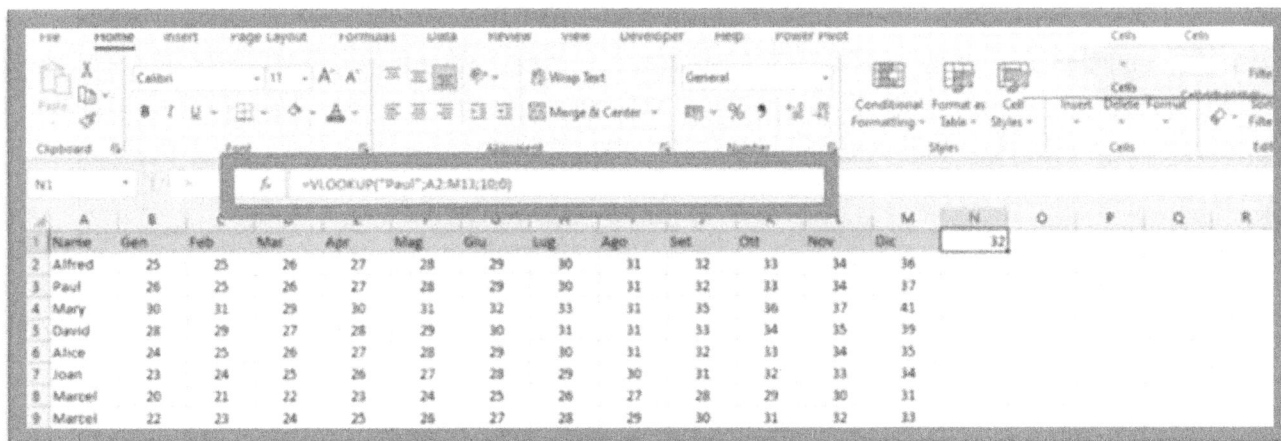

N1 f_x =VLOOKUP("Paul";A2:M13;10;0)

Name	Gen	Feb	Mar	Apr	Mag	Giu	Lug	Ago	Set	Ott	Nov	Dic	M	N
														32
Alfred	25	25	26	27	28	29	30	31	32	33	34	36		
Paul	26	25	26	27	28	29	30	31	32	33	34	37		
Mary	30	31	29	30	31	32	33	31	35	36	37	41		
David	28	29	27	28	29	30	31	31	33	34	35	39		
Alice	24	25	26	27	28	29	30	31	32	33	34	35		
Joan	23	24	25	26	27	28	29	30	31	32	33	34		
Marcel	20	21	22	23	24	25	26	27	28	29	30	31		
Martel	22	23	24	25	26	27	28	29	30	31	32	33		

This is the syntax of the formula, let's analyze it together:

= VLOOKUP ("Paul"; A2: M13,10,0).

Paul is the salesman whose work results we want to know in September. The range A2:M13 represents the table matrix. 10 is the row number that corresponds to September and 0 indicates the logical reference which means false.

Exercise: Using the same database, now go and look for Mary's sales in November.

When you want to locate objects in a table or range by rows, use the Vlookup command.

Choose a cell.

Enter =VLOOKUP (, and now choose the value you'd like Excel to search up.

After that, type a comma (,) and select the table or range where you wish to look for that value.

Enter a comma (,) followed by the column number where the lookup values are stored.

To locate the same match, type, FALSE).

Press the Enter key.

The VLOOKUP function looks like this;

=VLOOK (A7, A2:B5, 2, FALSE).

The latest version of the VLOOKUP function requires the input of a lookup value, the range containing the lookup value, the column number within the range that contains the output value, and the optional specification of TRUE for an approximate match or FALSE for an exact match.

By using the VLOOKUP function, you can combine data from multiple tables into a single worksheet.

The VLOOKUP function can be used to combine as many tables as possible into a single table, provided that one of those tables shares fields with all of the others. This is especially useful when sharing a workbook with people who are using old Excel versions that don't actively support data features with multiple tables as data sources. Simply combine the data sources into a single table and replace the data features' data sources with the new table. After you've completed this, you'll be able to use the data features in older versions of Excel (provided the data feature itself is supported by the older version).

Copies of the common fields table should be pasted into a new worksheet. The table should be given a name.

Click Data, then Data Tools, then Relationships to open the Manage Relationships dialog box.

For each relationship listed, keep the following in mind:

- The field that connects the tables (shown in parenthesis in the dialog box) is the LOOKUP value for the VLOOKUP formula.
- The Related Lookup Table is called: This is the table array in the VLOOKUP formula.
- The Related Lookup Table field (column) containing the required data in the new column: This information will not be visible in the "Manage Relationships" window because you will need to consult the Related Lookup Table to determine which field you want to obtain.

Fill in your VLOOKUP formula inside the first empty column with the information from step 3 if you want to add an extra field to the new table. Keep on adding fields till the needed fields you require are achieved. If you wish to create a workbook using data features that make use of several tables, just update the data feature's source of data to the new table.

HLOOKUP

Similar to VLOOKUP, another method named HLOOKUP(), or horizontal lookup, is available. The function HLOOKUP searches the top row of a table or array of benefits for a value. It returns the value in the specified column from the specified row.

The following are the HLOOKUP function's arguments:

- Lookup value — This specifies the value that should be looked up.
- Table — This is the table from which data must be retrieved.
- Row index — This is the row from which data will be retrieved.
- Range lookup — [optional] range lookup. This is a boolean value that indicates whether a match is precise or approximate. The default value is TRUE, which indicates a close match.

Consider the following table to demonstrate how to locate the city of Jenson using HLOOKUP.

	G	H	I	J	K	L	M
First Name	Ben	Stuart	Jenson	Lucy	Trent	Jhonny
Last Name	Zampa	Carry	Button	Davis	Patinson	Evans
Department	HR	Marketing	Operations	Sales	IT	Sales
City	Chicago	Kansas	New York	Los Angeles	Boston	Houston
Date Hired	10-11-2001	20-06-2002	01-12-2004	25-02-2011	17-08-2015	10-01-2018

Hlookup	
First Name	Jenson
City	=HLOOKUP(H23,G1:M5,4,0)

Hlookup	
First Name	Jenson
City	New York

Here, H23 represents the lookup value, i.e., Jenson, G1:M5 represents the table array, 4 represents the row index, and 0 represents an approximation match.

New York will display as soon as you hit enter.

XLookup

With the new Suite 365, Microsoft has released an update and a new Search function. Soon the XLOOKUP will replace the very popular vertical search because of its simplicity. With this function, we can find any element we need in a database inserted on an Excel spreadsheet. Even if not everyone is a subscriber to the 365 suite, knowing what's new in Excel can be useful.

Let's take a look at what this is all about.

Why should we use the XLOOKUP function?

I will only mention a few reasons:

1.   You can use a single formula to search for multiple items.

2.   You can search vertically, horizontally, in a database on the same worksheet, or even on another page. Don't you think it's convenient?

3.   Multi-line simultaneous search with multiple criteria is possible.

4.   It makes use of the wildcard character *

5.   You can look for exact matches, as well as major and minor variations.

In other words, this new function replaces and incorporates the previous Search family functions.

Now, let's look at the function syntax together. As we all know, knowing the exact syntax of the functions limits the possibility of receiving an error message.

=XLOOKUP(Value, Array, Return Array, [if not found], Match Mode, Search Mode).

The parameters surrounded by square brackets are optional.

•   Value: the item to be searched for can be numeric, text, or a wildcard character.

•   Array: the range in which Excel should conduct our investigation.

•   Return Array: The range in which the search result must be returned by Excel.

•   If not found: If Excel cannot find the requested value, we can enter what the program must release; if no indication is entered, the string ND is released.

•   Match mode: If necessary, we can specify a comparison mode in this section of the function. This is also an option.

•   Search mode: with this option, you can specify the starting point of the search, for example, 1 or -1, implying that we are only interested in positive numbers or, conversely, that we are also interested in that> 1. This string is optional as well.

In this function, there are three optional strings, this indicates the enormous possibilities that the function itself offers. Just think of the match mode; in addition to looking for a value that is textual or numeric, the search x is able to replace the compare function as well.

Chapter #11: Names and comments

How to Use a Named Range

To select a named range, click the dropdown arrow of the name box and select the name from the dropdown list. This will display the worksheet with the range (if you're on a different worksheet) and select all the rows and columns in the range.

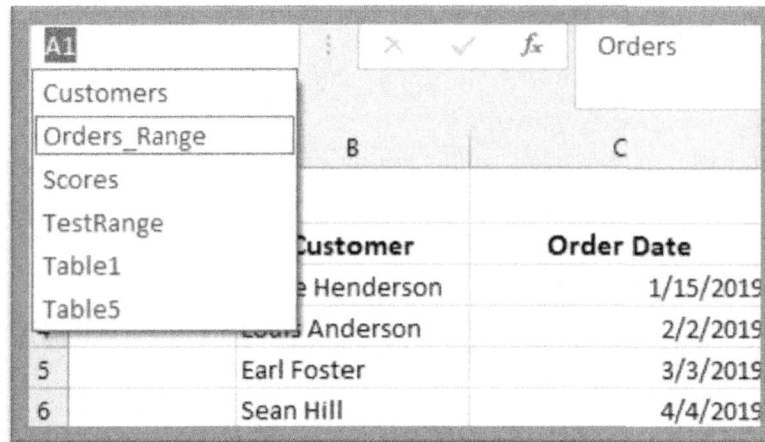

For example: The following example demonstrates the use of a named range called **Orders_Range** in place of the cell reference A1:D13. The example uses two formulas to count numeric values and blank cells in the range. The name of the range has been used as an argument in the functions instead of A1:D13.

=COUNT(Orders_Range) =COUNTBLANK(Orders_Range)

Excel Names Make Formulas Simpler to Re-Use

When you use Excel names, it is much easier to transfer the formula to another sheet or to a new workbook. Create a new workbook with the same name and copy/paste your formula: You'll be up and running quickly.

To prevent Excel from generating new names for this formula, copy the formula as text in the formula bar rather than a formula cell.

Navigation Named Ranges

To quickly access a specific range, click on its name in the name box. If a specified range is located on another document, Excel will automatically navigate to that sheet.

Complex named ranges in Excel won't appear in the name box. To see the dynamic ranges, open the Excel name manager (Ctrl+F3). It displays all of the names in the workbook, including their scope and references.

Ranges with Names Allow the Creation of Dynamic Drop-Down Lists

1. Create a complex named range first, and then rely on it to create an expandable and updatable dropdown list.
2. Getting Rid of a Named Range
3. Click Name Manager on the Formulas tab.
4. From the list, select the named range you want to delete.
5. Select the Delete option.
6. When you're finished, click Close.

How to Obtain a List of the Names in the Workbook

To create a much more meaningful list of your names in a new workbook, follow these steps:
1. Choose the topmost cell in the range to display the names.
2. Tap used in the formula and then paste names, then define name group in the formula tab. Alternatively, press the F3 key.
3. Paste names dialog box, select pastelist.

Both Excel names and references will be inserted into the current worksheet beginning in the selected cell.

How to Add Names to Existing Formulas

If you specify a range that is already included in the formulas, Excel will not immediately change your references to a proper name. You can make it instead of manually changing references with addresses. This is handled by Excel. This is how you do it:
1. By selecting one or more formula cells, you can upgrade them.
2. In the formulas tab, define the name group, press define name, and apply names.
3. In the Apply Names dialog box, select the name you want to use, then click OK. Excel compares a subset of the current names to the references in the formulas, but the names are chosen automatically. There are also two other options (both of which are selected by default):

When you want Excel to add only the names of the same relation type, leave the box checked: Ignore relative/absolute and replace relative references with relative terms, and absolute references with absolute names. If this option is selected, Excel will rename all cell references that are defined as the intersection of any named row and named column. For more options, click the options button.

Excel Shortcuts for Names

The most frequently used Excel tools can be accessed via the ribbon, the right-click menu, and keyboard shortcuts. Excel-named ranges are no exception. Here are three Excel shortcuts to help you work with names:

- Ctrl+F3 will bring up the Excel name manager.
- Ctrl+Shift+F3 to create a named range from a set.
- F3 will bring up a list of all the names in the workbook.

Name Error in Excel (#REF and #NAME)

When you insert and remove cells from a named set, Microsoft Excel does its best to keep the names consistent and accurate by changing the range references automatically. If you construct the named range from cells A1:A10 and then insert a new row somewhere between rows 1 and 10, the range reference changes to A1:A11. Similarly, deleting every cell from A1 to A10 shrinks the named range.

If any of the cells in a named range are removed, the name becomes invalid, and the name manager displays a #REF! Error. A similar error will appear in a formula that contains a reference to that name: A #NAME?

When a formula is applied to a name that does not exist, an error occurs (due to a typo or deletion). In any case, launch the Excel name manager and verify that the names you've identified are still correct (this is the fastest method for filtering names with errors). That's how you make and use names in excel.

Chapter #12: Pivot Tables

When it comes to Excel, pivot tables are an absolute must-know. They enable you to convert a dataset into a table, analyze data, apply filters, create charts and interactive dashboards, and much more.

Pivot Tables are a powerful tool that can count, sort, calculate, and summarize data for you automatically. They have advantages and disadvantages; they can be very useful, but they are not intuitive, and if you do not master them, they can slow down your work and waste time.

We will now look at Pivot Tables to learn the fundamentals, how they work, and see multiple strategies to help you use them to your advantage and save hours of work every time.

Transform Your Dataset into a Table

Every Pivot Table is linked to a data source. Every time your data source change (in this instance, our table with 7 customer records), you will have to manually *REFRESH* the pivot tables linked to it or *CHANGE DATASOURCE,* in the event your new data has more records or you simply want to place the data source in another sheet.

Transform Your Data Set into a Table

Select the entire table

- *SHORTCUT:* Click CTRL + T

Your selected data is highlighted in the data range box. Don't edit it. Just make sure all the data in your table is selected properly. Press Enter to confirm

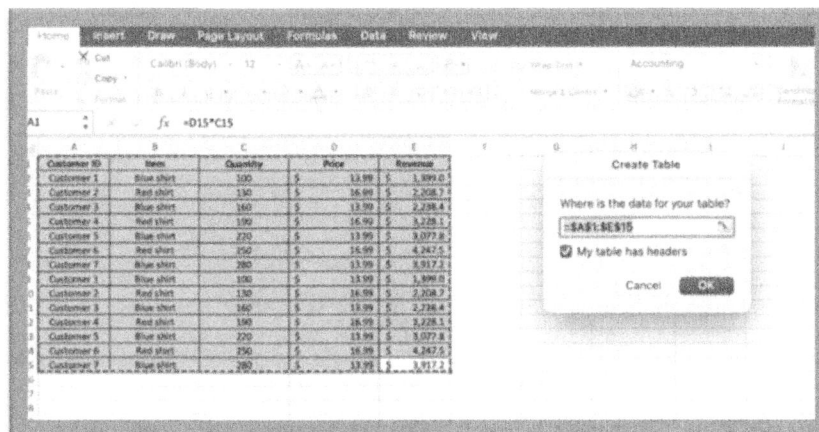

You can get the same outcome by selecting the table and press from the Ribbon
Home > Format as a Table

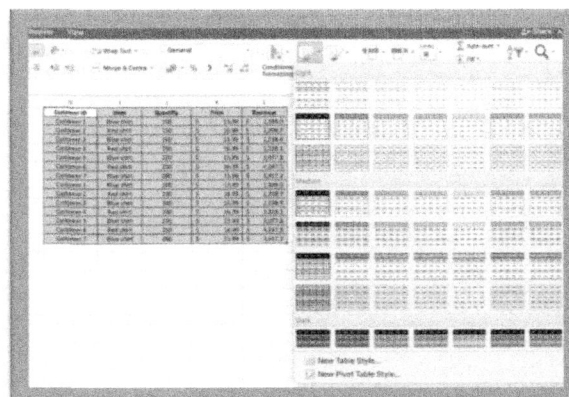

Once you format your data set as a table, you will see that the look of your data set changes.

You can click anywhere in the table at any time and you will notice that a new section named Table has been added to the Ribbon. From this section, you are able to change the table settings, select or create a style, change rows and columns, and much more.

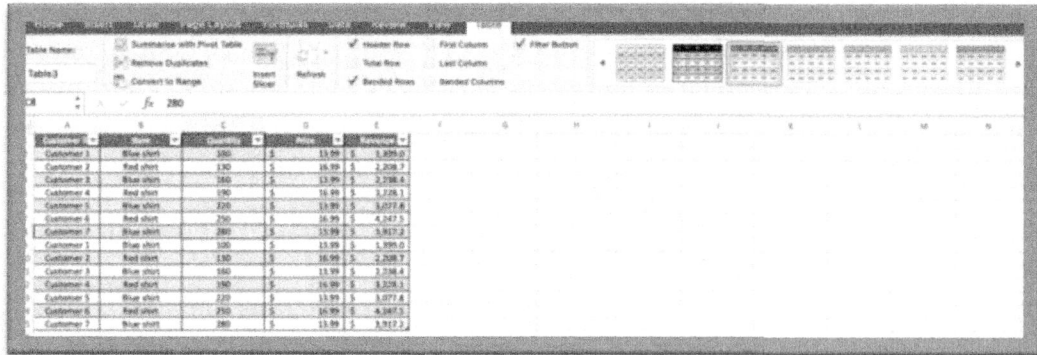

QUESTION: Why do you need to format your data as a table before creating a Pivot Table?

ANSWER: To automate the process. In fact, you will now choose your table to create a Pivot Table. This is not the same as creating a Pivot Table from a simple data range.

The issue is that all of the pivots you will create will pull data from the current range of your table, which is A1:E15.

But what if the range changes tomorrow to A1:E16, with a new row added to the bottom?

The Pivot Tables you created today will not update automatically, and you will need to manually update them all with the new data source to see the changes in the Pivots.

However, if you transform your data into a table BEFORE creating a Pivot Table, you will never have this issue, and the Pivot Table will automatically receive range updates as your data changes.

Multiply this by the number of data sets you will analyze, the number of pivot tables you will create, and the number of times you will need to update each table. It takes an incredible amount of time and manual labor.

This straightforward procedure will save you all of that time!

Now, you can even add a new row manually on the bottom and you will notice that it will become part of the table automatically.

Pivot Table Area

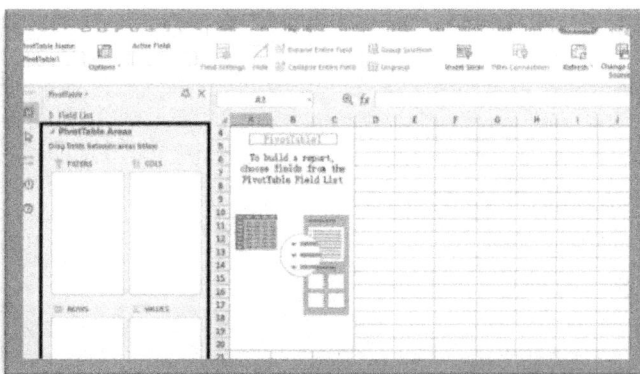

The Pivot table is where all the data that was selected from the Pivot field list is put together to make a table. The Pivot table has headings for each of the categories, and where each field for that category is placed in order under those headings.

There are four parts to a Pivot table, including the Filter area, Rows, Columns, and Values.

The Filter Area

The filter area appears on the right-hand side of the table and it is used to apply a condition to the data. For example, clicking the "Product A" filter in the first pivot table allows you to select a different Product range. The formula bar at the very top of Excel has a drop-down menu for automatically filtering data.

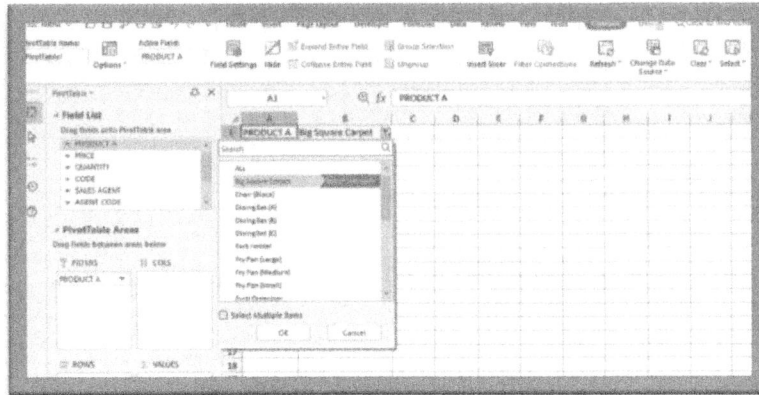

The Rows Area

The Rows area of an Excel spreadsheet shows the source data from which the Pivot table has been extracted. For example, the first column in the rows area shows the names of all the Sales Agents of the Product.

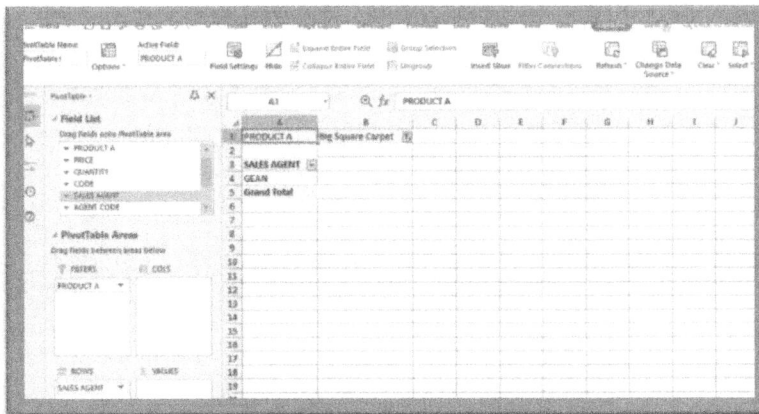

In the above example, the name "GEAN" appears under the row, it's because she's the sales agent of the said Product.

Now, let's filter the Product and select "ALL."

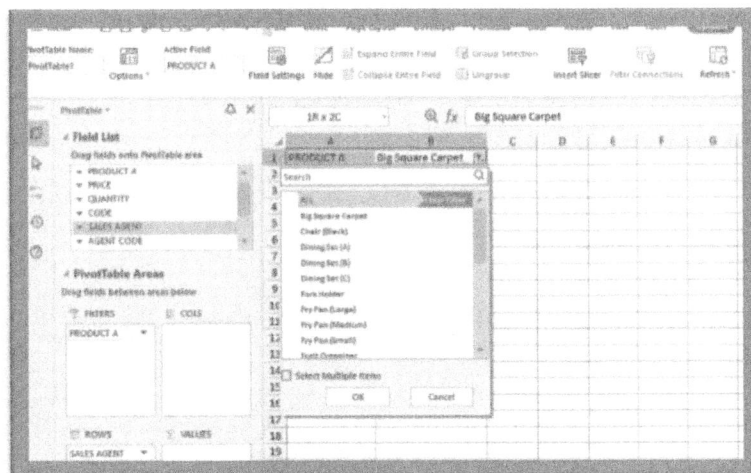

This is what it looks like:

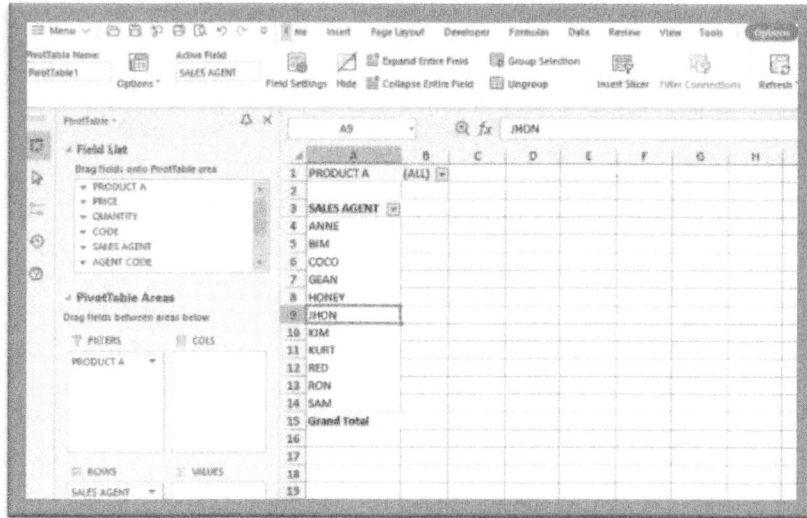

As you can see, all the name of the agents from the Product Range appears in our Pivot table. This is how to get information into your Pivot table.

The Columns Area

The columns of the Pivot table are arranged to show the information most meaningful to you. For example, in the second pivot table, the columns are arranged to show the Sales Code.

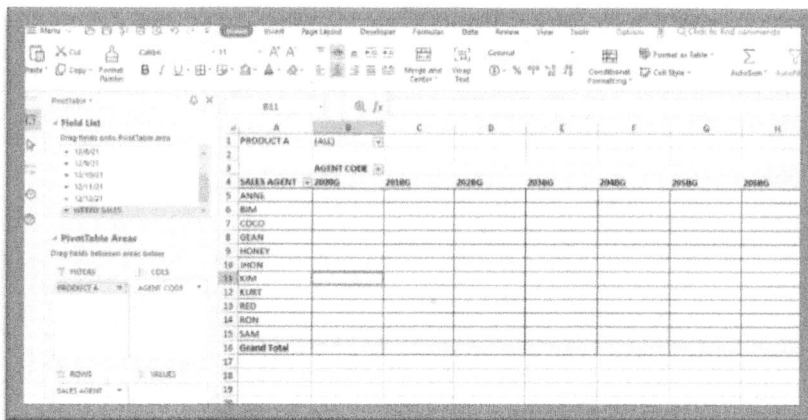

The Values Area

The Values Area is located at the bottom of a Pivot table and contains summary calculations created from the source data and can be used to further manipulate and analyze your worksheet data.

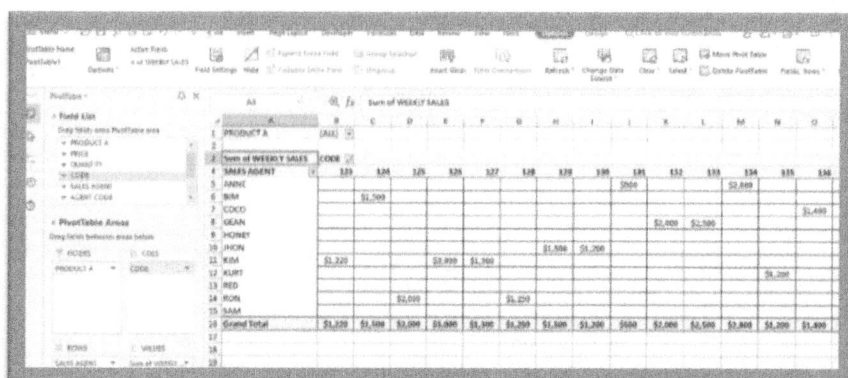

This is what your Pivot Table looks like

Using the Commands to Accomplish a Pivot Table

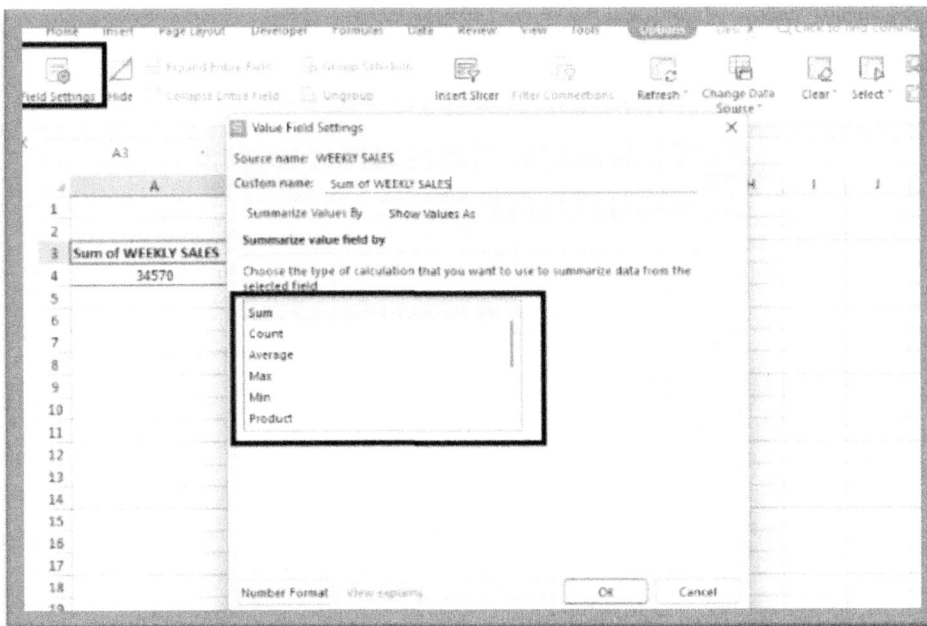

Pivot tables, unlike other spreadsheets, are not constrained by row and column dimensions. They can also have a value field with multiple dimensions. To determine which cells in column A will be evaluated for values, you must specify the settings for the value field and the value label. To do so, open the design tab at the top of your pivot table and then click the value field settings symbol beneath the fields.

You will then see the following settings:

There are 11 different operations that you can choose from. Depending on your requirement, these are SUM, COUNT, AVERAGE, MAX, MIN, PRODUCT, COUNT NUMBERS, STDDEV, STDDEVP, VAR, and VARP.

• Sum: This operation will summarize the values in a field or column as a numerical value.

• Count: This operation will count the number of values in a column or field and put it into another column.

• Average: This operation will calculate the average of values from a specified row or column and place it in another cell.

• Max: This operation will display the maximum value from a specified row or column and place it in another cell.

• Min: This operation falls on the minimum value from a specified row or column and places it in another cell.

• Product: Used to calculate the product of numbers that are found in different columns or rows, this option is selected if you want to multiply all the numbers that you have on your spreadsheet.

• Count Numbers: This operation will count the number of numbers in a column or row and put it into another column. This is helpful if you want to display only a certain number of numbers both in your cell and on your pivot table.

• STDev: is applied to calculate the standard deviation, which is a measure of how far the average value of a data set differs from the mean value.

• STDVEP: This operation turns out when you have calculated the standard deviation, this will calculate the value of the standard deviation and place it into another field.

• VAR: This operation calculates the variance from numbers in different columns or rows.

• VARP: This value returns the variance of numbers of the population. This is usually used in the context of being a function that is performed on a population.

Create a Pivot Table

1. Click anywhere in the Table
2. From the ribbon, click on Insert > Pivot Table
3. Press OK to confirm the creation of a Pivot Table in a new Worksheet (here named Sheet 2, it follows the number of the previous sheet)

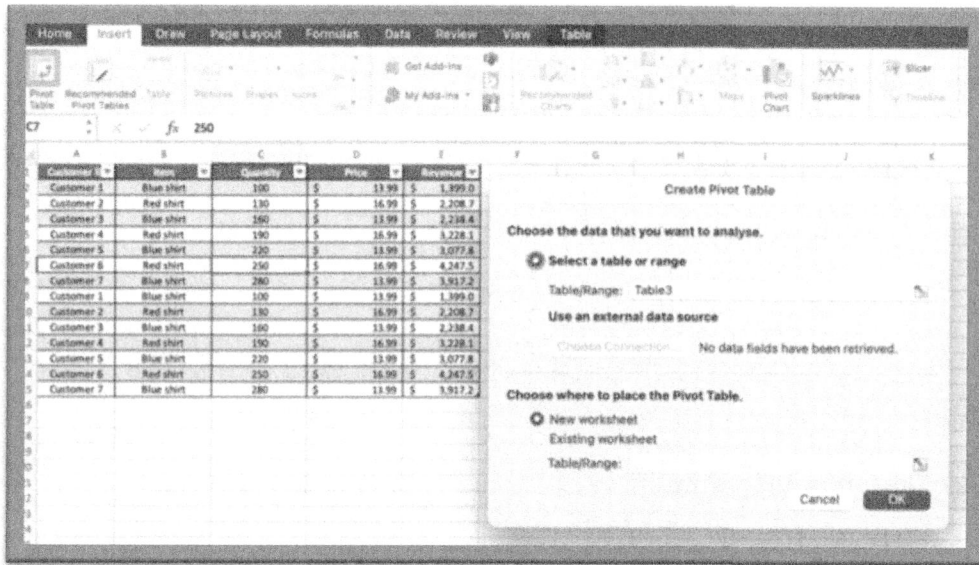

You can also add the pivot table to the current worksheet or CTRL + X, CTRL + V.
As you can see, a new sheet has been created.

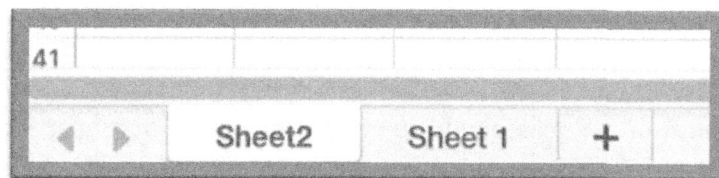

The new sheet has also 2 new sections in the Ribbon that will appear when you click on the Pivot Table. These sections are named Pivot Table Analyse and Design. You can already see the two buttons *REFRESH* and *CHANGE DATA SOURCE* we mentioned earlier.

You will need to refresh your data from the pivot if you update Sheet 1.

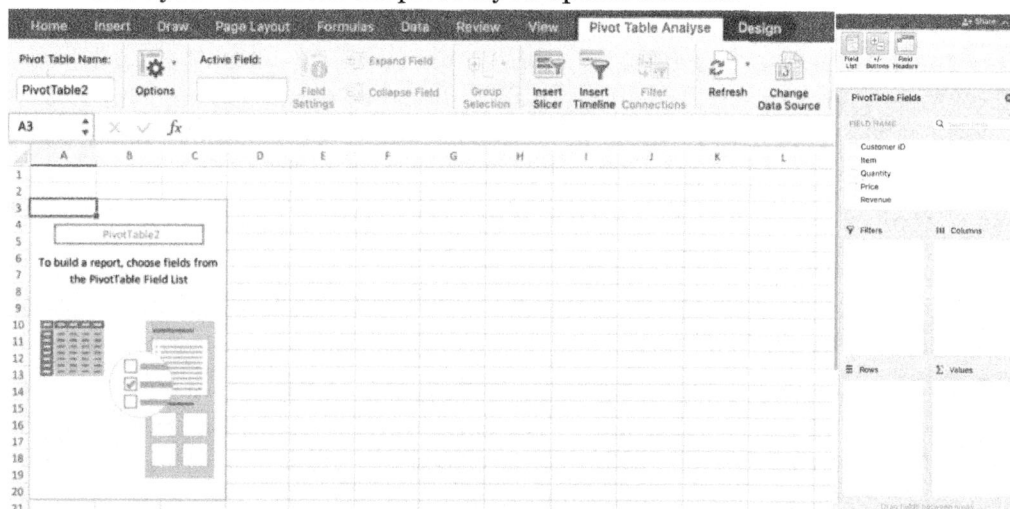

You may also notice the field list on the right side of the screen as well as four boxes (Filters, Columns, Rows, Values) that you will use to change the fields in your Pivot Table by dragging and dropping them. NOTE: If you create two separate pivot tables, you can access the linked field list and edit the table one at a time by clicking on each of them individually.

- IMPORTANT: This is the field list. It appears when you click on the pivot table and disappears when you click out. If, for some reason, you can no longer see it despite selecting the Pivot Table:

1. Place your cursor anywhere in the pivot table.
2. Select Pivot Table Analyze from the Ribbon.
3. Select Field List.
4. The Field List should have now re-appeared.

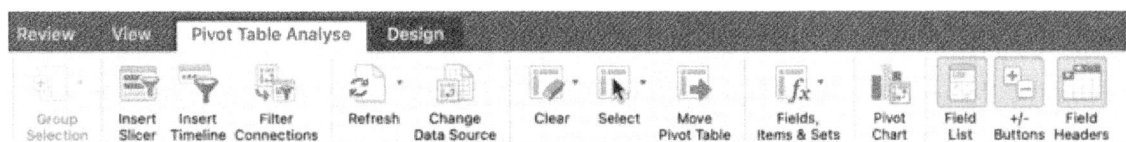

Now, let's start creating the table by dragging and dropping fields in the boxes.

1. Drag & drop Client ID in Row
2. Drag & drop Quantity in Values
3. Drag & drop Price in Values
4. Drag & drop Revenue in Values

This starts to look like the table we created at the beginning, but now the data is aggregated. Remember when we duplicated the table and pasted the values on the bottom?

This pivot table has all the customer values, but added up and summarized.

Formatting Tables

An Excel table can be formatted in numerous ways. There are pre-installed table styles in light, medium, and dark colors. Other table styles and options are listed further down.

Certain formatting will be applied in the following steps:

1. Locate the Excel ME sheet in your file.

2. In the table style group of the table tool design tab, press the further button.

Table Designs

1. Select table styles medium 7 from the medium table style gallery.
2. On the ribbon, select banded rows from the table style options group.
3. The contrasting color line fades away. The data in the table will become more difficult to read.
4. Experiment with any of the remaining table styles. Check the banded rows, header row, and filter button when you're finished.

Alter Pivot Table Source Data

Clicking on the pivot table you wish to alter will bring up the list of tools for use with the pivot table. Under the Data tab, select the Analyze option followed by Change Data Source.

In the Table/Range box, select the new range you want to use. Instead of typing it, simply select it on your worksheet, and it should auto-populate this section.

If your external data source has changed, you can see it by selecting the external data source option from the same menu.

Data model-based pivot tables cannot be changed.

Pivot tables, like regular tables, can be refreshed by clicking on the pivot table to bring up the tools for using the pivot table, then selecting the Data tab, followed by the option to Analyze, then Refresh or Refresh all to refresh all pivot tables in your workbook at once. You can also use ALT in conjunction with F5.

When altering data, ensure you select the prevent columns and cells from reformatting incorrectly by first selecting the Data tab, followed by the option to Analyze and then Options. Select the tab labeled Layout and Format and ensure that the options for column width and cell formatting are selected.

Filter and Sort a PivotTable

On some occasions, you may want to limit what is displayed in the PivotTable. You can sort and filter a PivotTable just like you can do to a range of data or an Excel table.

1.   To filter a PivotTable: Click on the AutoFilter (down arrow) on the Row Labels cell.

2.   The pop-up menu provides a list of the row headings in your PivotTable. You can select/deselect items on this list to limit the data being displayed in the PivotTable.

3.   *Uncheck* Select All.

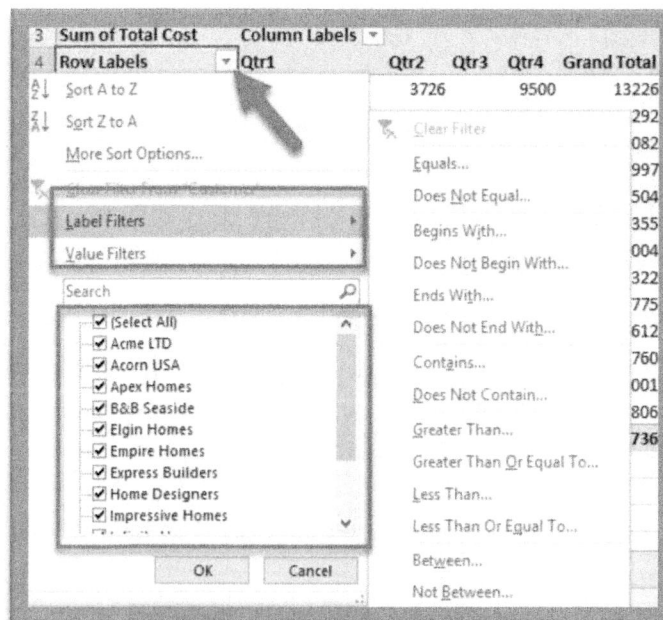

4.   Scroll through the list and manually select the items you want to display.

5.   Click *OK*.

6.   The PivotTable will now only display the columns you've chosen.

Using a Custom Filter

You can also apply a custom filter to your PivotTable by using the Label Filters and Value Filters menu commands. This is done in the same way that a range or table is.

Data Sorting in a PivotTable

To arrange your data in a PivotTable, use the same sorting methods as you would for a range or table. On the Row Labels column, click the AutoFilter button.

Sort A to Z (to sort in ascending order) or Sort Z to A (to sort in descending order). If your column headings are dates, you will get Sort Oldest to Newest (for ascending) and Sort Newest to Oldest (for descending).

Changing the Format of the Pivot Table

The style of a report can be easily changed. Anywhere in the book is a good starting point for clicking. On the Ribbon at the top of the screen, click the Design button under PivotTable Tools. To see all of the styles, click the More button in the PivotTable Styles group. When you move the pointer over a style, it appears in a temporary overview in the report.

To see all of the current models, scroll down. If you find one you want, press it to apply it to your pivot table.

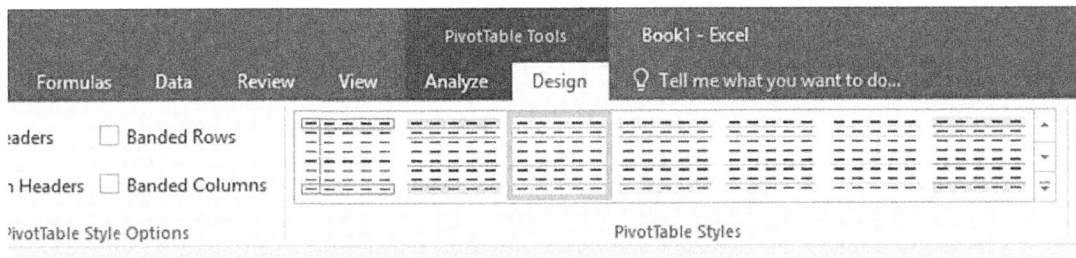

To test the choices, choose the *Banded Rows* and *Banded Columns* checkboxes. To remove banded rows or columns, remove the checkbox after selecting them.

The *Row Headers* and *Column Headers* checkboxes in the PivotTable Style Options group can also be used to switch off and on row and column header formatting. Press the dropdown arrow within the *Style* group and choose how you'd like the pivot table to look to attach blank rows or adjust the layout. Changes to the pivot table layout may be made under the *Design* tab, but they would have little effect on the data.

Creating a Pivot Table Chart

To generate a chart from the pivot table, go to the PivotTable Analyze section and select Pivot Chart from the Tools community. In the Insert Chart dialogue box, select the Chart you want to create, then press Ok. After you've created a chart from the pivot table results, the Pivot Chart tab will appear, and you'll be able to format it. To view all of the options, click the More button in the Chart Styles group on the Design tab, then select the type you want. Making a chart from a Pivot Table has the advantage of being interactive, allowing you to filter data from within the panel. In the class illustration, you have the option of sorting the Word and Subject.

Formatting a Pivot Table Chart

Using Add Chart Component feature under the *Design* tab, you can add names, axes, and gridlines to the Chart. Go to the *Chart Layouts* tab and then to the *Attach Chart Element* tab to attach a Chart Title. The choice to add a chart title is available here. There are various choices for formatting the PivotTable Chart under the *Format* tab.

Adjust the color and type of the text in the Pivot Table Chart by going to the *Format* tab and selecting WordArt Styles. You may even apply outlines to the content in the Chart by going to the *Format* tab and selecting WordArt Styles. Inserting a slicer into a Pivot Table map is also an alternative. Select the *Analyze* button from the *Pivot Chart* tab. Pick Insert Slicer from the Insert Slicer icon. Click *Well* after selecting the data type you want to use as a slicer.

Refreshing Excel Pivot Table Objects

The Pivot Table object is only as good as the data it is based on, which can change frequently. Adjustments at the source level may not be reflected if the pivot tables are included in the dashboard. Depending on the data in the pivot tables, refreshing them may be critical.

How to refresh:

1.   When you open the workbook, you should refresh it.

2.   After making changes to the underlying data, users may forget to reload. Furthermore, if pivot tables are linked to an external source, you'll need current values. The simplest way to ensure that users receive the most up-to-date information is to require that the pivot tables will be updated when you open the file in Excel.

3.   When you right-click any pivot table, select the PivotTable Option from the submenu that appears.

4.   In the resulting window, click the Data tab.

5.   When you open the file, select the option to Refresh data.

6.   To confirm the modification, click *Ok*.

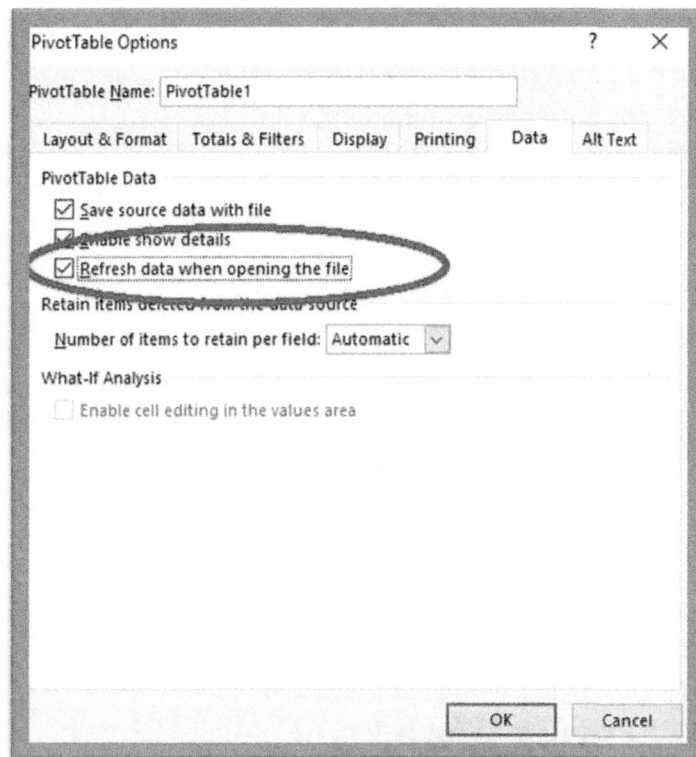

If you check this box, the chosen PivotTable (or all the pivot tables with the same data sources) will be updated. Set the options for all pivot tables that have different data sources if you have more than one.

PART THREE: ADVANCED LEVEL

Chapter #13: An Eye for Advanced Functions

Information Functions

- ISBLANK
- ISERROR
- ISNUMBER
- ISFORMULA

These functions are explained below step by step.

ISBLANK
ISERROR

Syntax	= ISERROR (value)
Return value	A logical value (TRUE or FALSE)
Arguments	The meaning will be checked for any errors.
Purpose	Checking the value

The ISERROR function in Excel returns TRUE for every error type, including #N/A, #REF!, #VALUE!, #DIV/0!, #NAME?, #NUM!, and #NULL! When you use ISERROR with the IF function, you can check for errors and show a custom message or perform a different calculation if one is detected.

	C5		f_x	=ISERROR(B5)
	A	B	C	D
1		ISERROR (value)		
2		Test for any error		
3				
4		Values	Result	Notes
5		#DIV/0!	TRUE	
6		#NAME?	TRUE	
7		#N/A	TRUE	TRUE for #N/A, unlike ISERR
8		#REF!	TRUE	
9		#NUM!	TRUE	
10		#REF!	TRUE	
11		#VALUE!	TRUE	

Notes:

- To see whether a cell contains some error messages, such as #N/A, #VALUE!, #REF!, #DIV/0!, #NUM!, #NAME?, or #NULL!, use the ISERROR function.

- =ISERROR(A1), for example, will return TRUE if A1 is showing one of the errors listed above and FALSE otherwise.

- Value is usually provided as a cell address, but it may also be used to catch errors in more complicated formulas.

ISNUMBER

Syntax	= ISNUMBER (value)
Return value	A logical value (TRUE or FALSE)
Arguments	The value to examine.
Purpose	Check for a numerical value.

When a cell contains a number, the Excel ISNUMBER function returns TRUE. Otherwise, it returns FALSE. ISNUMBER may be used to verify that a cell includes a numeric value or that a function's output is a number.

Use the ISNUMBER function to determine whether a value is a number. ISNUMBER returns TRUE when the value is numeric; otherwise, it returns FALSE.

=ISNUMBER(A1) would return TRUE if A1 contains a number or formula that returns a numeric value. If A1 contains text, ISNUMBER will return FALSE.

Notes:

• Value is typically supplied as a cell address, and the result can be evaluated by enclosing other functions and formulas within ISNUMBER.

• ISNUMBER would return TRUE for Excel dates and times and FALSE for numbers typed as text because they are numeric.

• IS functions are a group of functions that include ISNUMBER.

ISFORMULA

Syntax	= ISFORMULA (reference)
Return value	TRUE or FALSE
Arguments	reference — a reference to a cell or a range of cells
Purpose	Check to see if a cell has a formula.

The ISFORMULA method in Excel returns TRUE if a cell includes a formula and FALSE otherwise. ISFORMULA returns TRUE when a cell includes a formula, irrespective of the formula's error OR output conditions.

The ISFORMULA function can be used to check whether a cell includes a formula. If a cell includes a formula, ISFORMULA returns TRUE; otherwise, it returns FALSE.

You should use a keyboard shortcut to briefly highlight all calculations in a worksheet. Use the FORMULATEXT function to extract and present a formula.

Notes:

- In Excel 2013, the ISFORMULA function was added.

Logical Functions

Many advanced formulas depend on Excel's logical functions as a platform. The conditional values TRUE or FALSE are returned by logical functions.

- AND Function
- OR Function
- NOT Function
- IFERROR Function
- IFNA Function
- IF Function
- IFS Function

AND Function

Syntax	= AND (logical 1, [logical 2], ...)
Return value	TRUE if all premises evaluate TRUE; if not, FALSE
Arguments	Logical 1 — The 1st condition or logical value to examine. Logical 2 — The 2nd condition or logical value to examine.
Purpose	Evaluate multiple conditions with AND

The AND feature in Excel is a logical function that is used to combine several conditions at the same time. AND either returns TRUE or FALSE. Using =AND(A1>0, A110) to see whether a number in A1 is higher than 0 and less than ten. The AND feature, which can be paired with OR function, can be used as a logical test within the IF function to eliminate additional nested IFs.

The AND function can evaluate a large number of logical conditions at once, up to 255 in total. Every logical condition (logical 1, logical 2, etc.) must return FALSE or TRUE or arrays, references containing logical values.

AND evaluates all input values and returns TRUE only if they are all TRUE. If every value evaluates to FALSE, the AND feature will return FALSE.

Notes:

- The AND function is not very sensitive.
- The AND function does not support wildcards.
- Arguments based on text values or void cells are ignored.
- The AND function will return #VALUE if no logical values are identified or generated during evaluation.

OR Function

Syntax	OR (logical 1, [logical 2], ...)
Return value	If all of the arguments test to TRUE, TRUE; otherwise, FALSE.
Arguments	Logical 1 — The 1st condition or logical value to examine. Logical 2 — The 2nd condition or logical value to examine. (Optional)
Purpose	Evaluate multiple conditions with OR

OR is a logical function that can be used to evaluate multiple conditions at once. OR yields one of two results: FALSE or TRUE. For example, to calculate A1 for x or y. To prevent additional nested IFs, the OR function, which can be combined with the AND function, can be used as a logical measure within the IF function.

You can use the OR feature to measure multiple conditions at once, up to 255 in total. Each logical state (logical1, logical2, etc.) must either return FALSE OR TRUE or be sequences or references that contain logical values.

OR function will test all of the input values and return TRUE if all of them are TRUE. OR function would return FALSE if all logical evaluate FALSE.

Notes:
- Each logical condition must return TRUE or FALSE or be sequences or references of logical values.
- Arguments of text values or void cells are overlooked.
- When no logical values are detected, the OR feature will return #VALUE.

IFS Function

Syntax	=IFS (test 1, val 1, [test 2, val 2], ...)
Return value	The value corresponds to the first TRUE outcome.
Arguments	Test 1 — 1st logical test. Value 1 — Result when test 1 TRUE. Test 2, value 2 — Second value/ test pair [optional]
Purpose	Test several conditions and return the first one that is true.

The Excel IFS method runs multiple experiments and returns the first TRUE outcome as a value. To test different conditions without several nested IF statements, use IFS function. IFS allows formulas to be simpler and quicker to learn.

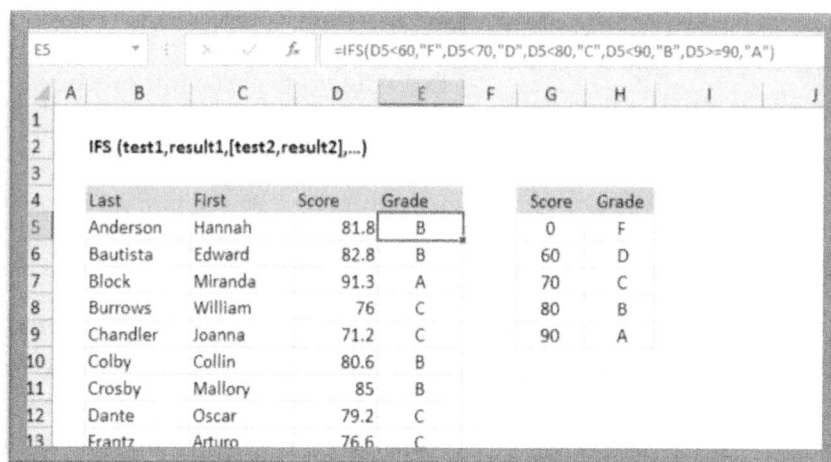

Notes:

- If all parameters are FALSE, the IFS feature does not have a built-in default value to use.
- Enter TRUE as a final test and value to return if/when no other conditions are met to have a default value.
- The results in all logical evaluations must be TRUE or FALSE. Every other outcome would result in a #VALUE! Error from IFS.
- IFS can return the #N/A error unless no logical tests return TRUE.

Index Function

The INDEX function, when combined with the MATCH function, can also replace the VLOOKUP function in the family of search and analysis functions.

The INDEX function returns the cell value that is included in the intersection of a row and a column in a given range.

It has the syntax = (Array; Row Num; [Col Num]).

- The Array represents the range in which we look for the value.
- The Row Num indicates the row within the matrix from which the value is returned.
- Col Num denotes the column from which the value is extrapolated.

Experiment by going to a price list type database and extrapolating the unit cost of an item contained therein.

Match Function

The MATCH function identifies the location of a value in the selected range.
Its syntax is = (Value; Array; [Match_Type]).
- Value is the value to search for.
- Array is the range in which the value is included
- Match_Type is the search criteria to be defined; exact match, greater or lesser. 0 will be indicated for the exact match; 1 for major correspondence; -1 for the smallest match.

IF Functions

	A	B	C	D
1	Data		Formula	Answer
2	2		=IF(A2>3,"yes","no")	no
3	2			no
4	5			yes
5	3			no
6	4			yes
7	2			no
8	3			no
9	3			no

The IF feature is frequently used when sorting data based on a set of logic. The IF formula has the advantage of allowing the use of formulas and functions.

Using this formula, Excel will tell you whether a specific criterion is met. For example, one might want to know which data in column A are greater than four. Excel will easily generate a "yes" for each cell with a value greater than 4 and a "no" for each cell with a value less than 4.

This is a well-known formula.
- =IF (logical test, [value if true], [value if false])

Take the following scenario:

- =IF (C2 is less than D3, 'TRUE,' 'FALSE') — If the value of C3 is less than the valuation at D3, the condition is true. If the rationale is right, set a cell value to TRUE; otherwise, set it to FALSE.
- =IF (A2 is greater than 3, "Yes," "No").

For example: =IF (B3="Yes", 1, 2), says IF (B3 is true, produce a 1, otherwise output a 2).

The "IF" Syntax

The "IF" function may be seen as a logical function that can return a value if a condition is true and returns another value if the condition is false.

IF (logical test, value if true, [value if false])

For instance:

- =IF (A2>B2,"Over Budget","OK")
- =IF(A2=B2,B4-A4,"")

Name of argument	Description
Logical_test	This means the condition you wish to examine
Value_if_true	This means the value you wish to return if the test is TRUE
Value_if_false	This means the value you wish to return if the test is FALSE

Easy IF Cases

	X ✓ _fx_	=if(c2="yes",1,2)		
C	D	E	F	
7				
9			:"yes",1,2)	
0				

=IF(C2="Yes",1,2)
In the example above, cell D2 denotes: If C2 is true, produce a 1, and otherwise output a 2.

	_fx_	=IF(C2="yes",1,2)		
	D	E	F	
7				
9			2	
0		no		

- =IF(C2=1, "Yes", "No")

In the preceding example, the formula in cell D2 is as follows: IF (If C2 is greater than one, output Yes; otherwise, output No.) As a result, it should be obvious that the "IF" function can be used to evaluate both values and texts. In order to evaluate errors, the "IF" function can be used. The "IF" function can be used to do more than just check if one term is equal to another and return a single answer; you can use mathematical operators to perform additional computations based on your needs. You can nest multiple "IF" functions together to perform multiple comparisons.

BESSELK		▼ X ✓ _fx_	=if(c2>b2,"over","within")			
A	B	C	D	E	F	G
udget	actual	status	amount			
54	64	over	10	"within")		
98	78	within	0			
87	67	within	0			
89	99	over	10			

- =IF(C2>B2, "Over", "within")

The IF function in D2 implies IF(C2 Is Greater Than B2, return "Over Budget," otherwise return "Within") in the example above.

	A	B	C	D	E	F
1	budget	actual	status	amount		
2	54	64	over	10	c2-b2,0))	
3	98	78	within	0		
4	87	67	within	0		
5	89	99	over	10		

- =IF(C2>B2,C2-B2,0)

Rather than returning a text result in the example above, mathematical computations will be returned. If you look at the formula in E2, you'll notice that it says IF (Actual is Greater than Budgeted, then Subtract the Budgeted from the Actual, else return nothing).

- =IF (E7="Yes",F5*0.0825,0)

The calculation in F7 in the preceding example implies IF (E7 = "Yes," then compute the Total Amount in F5 * 8.25%, otherwise no Sales Tax is owed, thus return 0).

The text must be enclosed in quotes if it is to be included in a formula. This rule does not apply to the use of TRUE or FALSE, which Excel understands.

Using the "If" function to determine if a cell is empty.

You may need to double-check if a cell is blank from time to time, perhaps because you don't know how to get an output without a tangible input.

You can use the IF with the ISBLANK function in this situation:

- =IF (ISBLANK (D2),"Blank", "Not Blank")

That is, IF(D2 is blank, return "Blank", otherwise return "Not Blank") You can also come up with your own formula for the "Not Blank" criterion. Instead of ISBLANK, we'll use "" in the below scenario. *The "" stands for "nothing."*

	fx	=IF(D3="","blank","not blank")						
C	**D**	**E**	**F**	**G**	**H**	**I**	**J**
item	quantity						
widget	2	$2.90	$5.80				
doohickey	3	$8.55	$25.66				
	sub total	$11.45	$31.46				
						not blank	
	sales tax		$2.60				
	total		$34				

- IF (D3="","Blank", "Not Blank")

If D3 is null, return "Blank," else "Not Blank," according to the above formula. A popular way to use "" to prevent your formula from calculating if a dependent cell is genuinely blank is as follows:

- =IF (D3="","",Your Formula())
- IF (D3 is nothing, then return nothing, else calculate your formula).
- IFS (Intelligent Function System)

The IFS function can be used to check if one or more criteria have been met, and then output a result that accurately matches the first TRUE condition. IFS can be used instead of numerous nested IF statements, and it is actually easier to read when there are multiple conditions.

IFERROR

The IFERROR function is not widely known or used, even among advanced Excel users, but it can be very useful when you need to deliver a job to a client and give the exact name of the error, rather than the generic errors that would make our work appear to be incorrect.

Let's return to our previous database with Name, ID, Address, City, and ZIP code. If we go to look for a name that is not in the list, Excel will obviously return #ND as an error message, but we want to make it clear to the customer that this is not our fault. We can tell Excel what to write if it can't find our query using a very simple syntax. It only works with structured errors and not with function setting errors. The function's syntax is IFERROR (value; value if error). In value if error, we could enter a name that isn't on the list or anything else we want.

Try this exercise to put your skills to the test: Return to the previously worked-on database and find the name Jim Morris, for example, and decide what wording you want to appear in the cell instead of #ND.

The idea that Excel is a difficult program to understand and use is a tiring If you use it frequently, it will become second nature.

Prepare a table based on the following criteria:

- Turnover — first quarter-second quarter-third quarter-fourth quarter
- Expenses — first, second, third, and fourth quarters

- Losses or revenues: these are the figures from the previous two tables.

Prepare a second page where you will record the receipts and expenses for each month.

Finally, go to format the currency in € and format the negative results in red. Save this table as we will use it again in the Charts chapter.

Offset Functions

The OFFSET function in Excel is very useful, it returns a reference shifted in a certain number of rows and columns. Its syntax is (reference; rows; cols; [height]; [width]). As always, the arguments enclosed in square brackets are optional.

- Ref: cell in which the movement is to be made
- Rows: to discard with a positive number move to the right, with a negative number to the left.
- Columns: number of columns to discard always using positive or negative numbers
- Height: maximum number of lines to consider. This argument is optional.
- Width: number of columns to consider. This argument is optional.

Chapter #14: Advanced Conditional Formatting

Conditional formatting is essential for displaying data in a worksheet or database. In practice, it is a graphical display based on a predefined criteria.

It is easily accessible from the Home Tab, which is also known as the Formatting Tab. Clicking on the Tab opens a drop-down menu where you can enter the criteria that can improve the performance of your worksheet, such as getting a graphic overview of the sales trend, checking the minimum and maximum sales, finding and deleting dubbed lines, and much more. Simply browse through all of the pre-set criteria to use this formatting.

Conditional formatting is especially useful in three scenarios:

1. Display values: gives you the opportunity to display the values you have selected in a different color, for example, sales below € 2000, or higher, or the general trend with handy arrows, or a gradient that allows you to color in different shades according to the criteria you entered. In addition, in the various options, you can also find a function that can facilitate your work when you need to take an overview of the company's performance.

2. Check data: in this case, conditional formatting helps you highlight errors, duplicates, or data that can slip by a not in-depth reading. The different coloring of the cells immediately highlights the differentiations you decide to give.

3. An important function of this type of formatting is precisely the ability to filter duplicated data or, on the contrary, unique data.

The Tab can be used to create conditional formatting, but there is a faster way to choose formatting. Simply select the reference cells, click on the formatting symbol, and a window will open, displaying all of the formatting rules you have selected. You can format your data based on certain conditions to display a visual representation that helps you spot critical issues and identify patterns and trends using conditional formatting. You can use visual representations, for example, to clearly show the highs and lows in your data as well as the trend based on a set of criteria.

In the example below, we can quickly see the trend in sales and how they compare to each other.

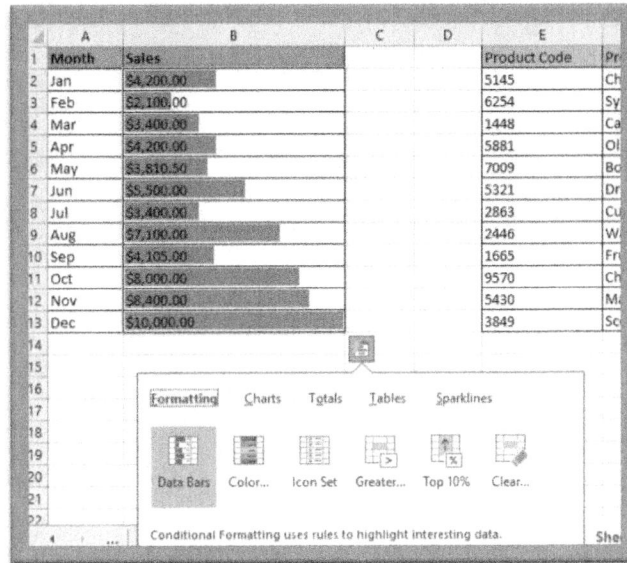

To quickly apply a conditional format:

1.   Select the range of cells you wish to format. The quick analysis button will be displayed at the bottom-right of the selection.

2.   Click the Quick Analysis button, and use the default *Formatting* tab.

3.   When your mouse hovers over the formatting options, you'll see a live preview of what your data will look like when applied.

4.   Click on *Data Bars* (or any of the other options) to apply the formatting to your data.

Use Multiple Conditional Formats

You may use multiple conditional formats on the same group of cells.

1.   To do so, select the cells, click the Quick Analysis button, and click another format option, for example, Icon Set.

2.   The arrows are used to depict the upper, middle, and lower values in the set of data.

Formatting Text Fields

For example, if we wanted to highlight all the rows with "Sauce" in the name, we would: Select the range.

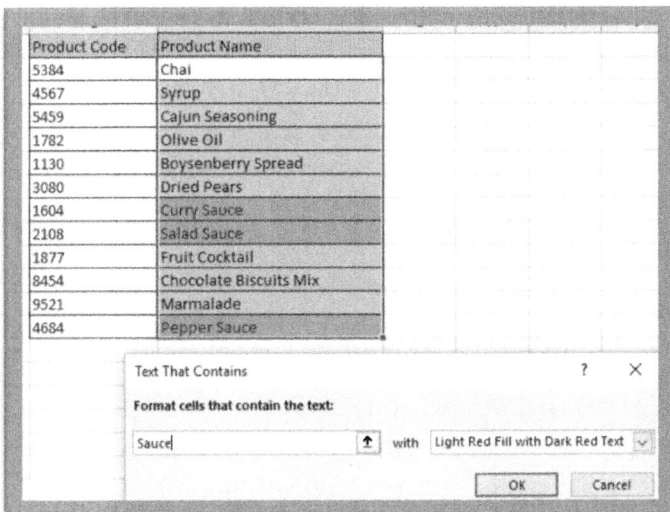

Click the Quick Analysis button.

Select *Text...* from the Formatting options.

In the text that contains dialog, we would enter sauce in the first box and select the type of formatting we want from the dropdown list.

You can explore the formatting options for different data types by selecting the data to be formatted and clicking on the *Quick Analysis* button.

Creating Conditional Formatting Rules

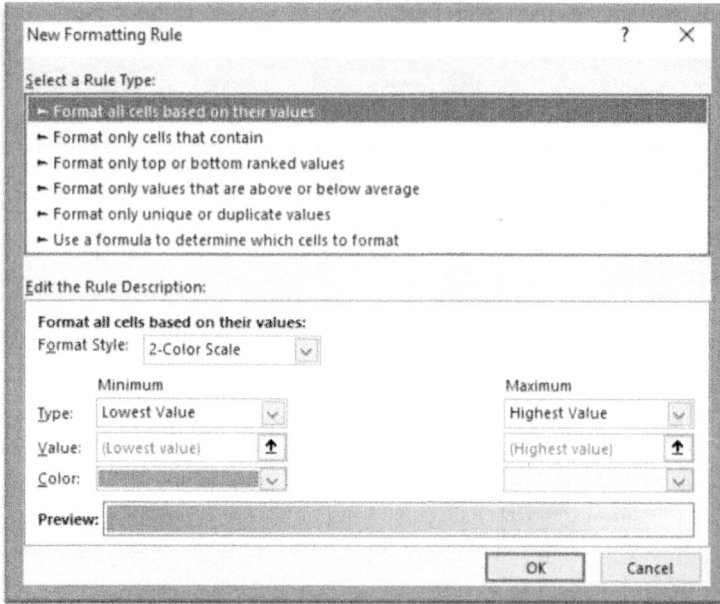

An alternative way to create conditional formatting is by creating Rules in Excel.

To launch the New Formatting Rule dialog: On the Ribbon, click on *Home > Conditional Formatting > New Rule*.

You can use this dialog to create more complex rules using a series of conditions and criteria.

You can choose a rule type from the options listed below:

- Format all cells according to their values.
- Only cells whose content should be formatted.
- Only the top or bottom-ranked values should be formatted.
- The bottom half of the screen, labeled Edit the Rule Description, gives you different fields to define your rule for each rule type.

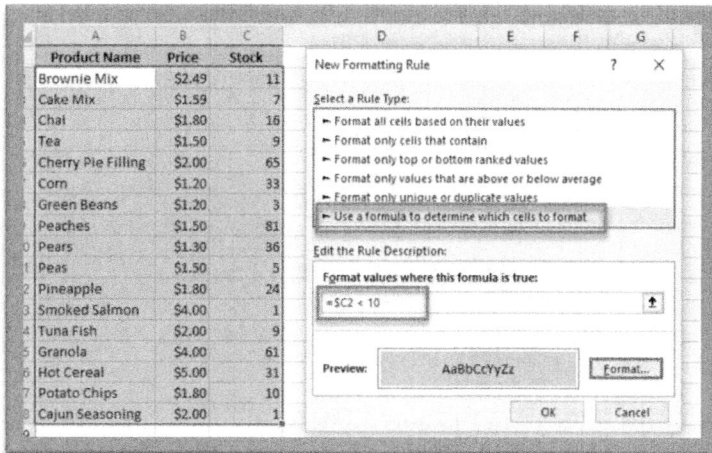

Example: Assume you have a products table and want to make the entire row grey if the product stock falls below 10.

To do this, select the range to conditionally format, such as A2:C18. It should be noted that A2 is the active cell.

Select Conditional Formatting > New Rule from the Ribbon. To determine which cells to format, use a formula.

Since A2 is the active cell, you need to enter a formula that is valid for row 2 and will apply to all the other rows.

To do this, you type in the formula =$C2 < 10. The dollar sign before the C means it is an absolute reference for column C ($C). With this, the value in column C for each row is evaluated and used for conditional formatting.

For the fill color, click the *Format* button, select the fill color you want and click OK and OK again to apply the rule.

The rows with Stock below 10 will now be filled with grey.

Carrying out Calculations with Formulas

Excel provides tools and features that enable you to carry out different types of calculations from basic arithmetic to complex engineering calculations using functions.

Conditionally Formatting Time

Let's say we had a task list, and we wanted to see which ones were late, i.e., the ones with the due date before today.

1.   You must choose the cells in the Due date column.
2.   Click the Quick Analysis button, followed by Less Than.
3.   Enter =TODAY (). We could type in today's date, but that would require us to update the conditional formatting on a daily basis, which could become tedious quickly! The TODAY function always returns the current date.
4.   Choose the formatting you want to use from the dropdown menu.
5.   Click *OK*.

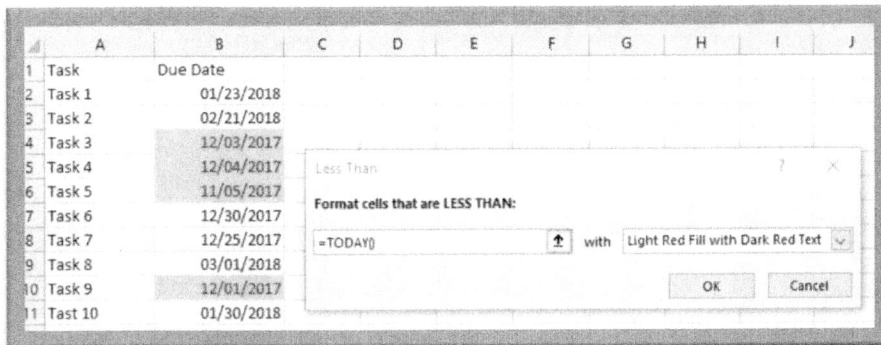

The tasks that are overdue now stand out in the list and are easy to identify at a glance.

Conditional Formatting and Table Format

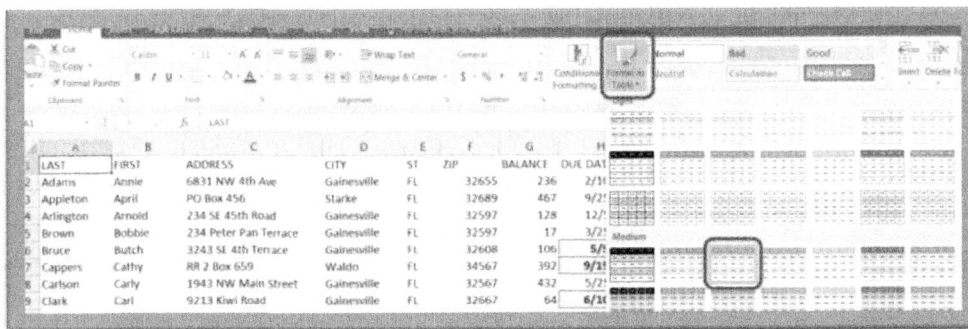

Using conditional formatting, you can highlight trends and patterns in your data. You can create rules to govern the format of cells based on their values as seen in the quarterly temperature readings with cell colors based on their values. Conditional formatting can be applied to any type of cell (either a selected range or a named range), an Excel table, or even a PivotTable report in Excel for Windows.

Quick Analysis is a function that allows you to use the Quick Analysis button to apply specified conditional formatting to the currently selected data. When you select data, the Quick Analysis icon appears on the screen right away.

Select the data that you want to conditionally format. The Quick Analysis icon appears at the bottom of the list.

To perform a quick analysis, click the Quick Analysis button or press CTRL.

The Formatting tab will appear in the pop-up window that displays. Move your cursor over the various formatting choices to get a Live Preview of your content, and then select the formatting type you wish to use.

Chapter #15: Advanced Filters

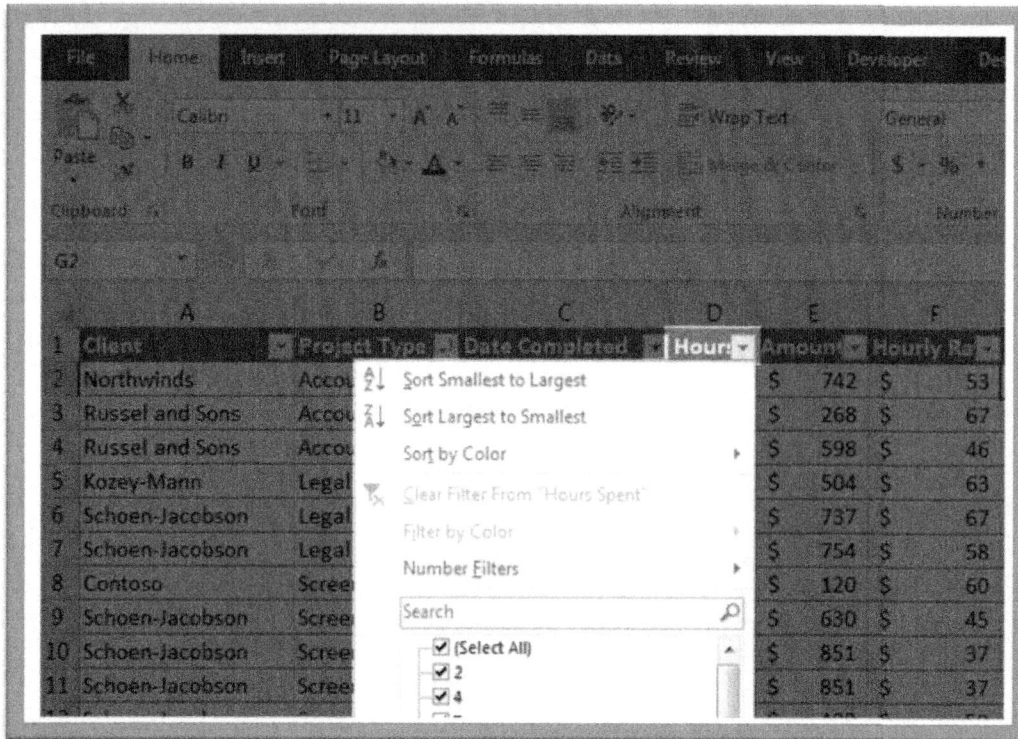

You may notice that when you convert data to a table in Excel, filter buttons appear at the top of each column. These provide a simple method of limiting the data that appears in the spreadsheet. To open the filtering box, click on the dropdown arrow.

You can check or uncheck the boxes for the data you wish to remove from the table view.

Again, this is a feature that justifies the use of Excel tables. While it is possible to implement filtering without using tables, with so many capabilities, it makes more sense to convert to tables.

Excel tables will have column headings by default; however, if your data is not an Excel table, ensure you have column headings like Category, Product Name, Price, etc. This makes using sort much easier.

Category	Product Name	Price	Reorder Level	Target Level
Beverages	Chai	18.00	10	40
Condiments	Syrup	10.00	25	100
Condiments	Cajun Seasoning	22.00	10	40
Oil	Olive Oil	21.35	10	40
Jams, Preserves	Boysenberry Spread	25.00	25	100
Dried Fruit & Nuts	Dried Pears	30.00	10	40

You can add column headings to your data by inserting a new row at the top of your worksheet and entering the headings. This is important because Excel will use the first row for the filter arrows.

How to Sort Data: Choose any cell in the data set that you want to filter. Select Home > Sort & Filter > Filter (or Data > Filter). Each column will have a filter arrow at the top. This is also known as an AutoFilter. It's worth noting that filter arrows are enabled by default in Excel tables.

To filter a column, click the AutoFilter button. Price, for example.

1.   Uncheck Select All and then check the values for the filter.
2.   Click *OK*.

The AutoFilter changes to a funnel icon to show that the column is filtered. If you look at the row heading numbers, you'll see that they're now blue, indicating which rows are included in the filtered data.

Applying a Custom Filter

Click on the AutoFilter of the column you want to use for the filter.

You'll see a menu item and a pop-out menu on the pop-up menu. Depending on the data type of the column, you'll have the following options: Text Filters are available when the column contains a text field or a combination of text and numbers: Equals, Does Not Equal, Starts, Ends, or Contains

Number Filters are only available when the column contains only numbers: Equals, Does Not Equal, Greater Than, Less Than, or Between.

Date Filters are only available when the column contains only dates: Last Week, Next Month, This Month, and Last Month.

Clear Filter from 'Column name' — this option is only available if the column has already had a filter applied to it. To clear the filter, select this option.

When you select any of the first 3 options, you will get a dialog box — *Custom AutoFilter*. You'll be specifying your custom filter conditions using this screen.

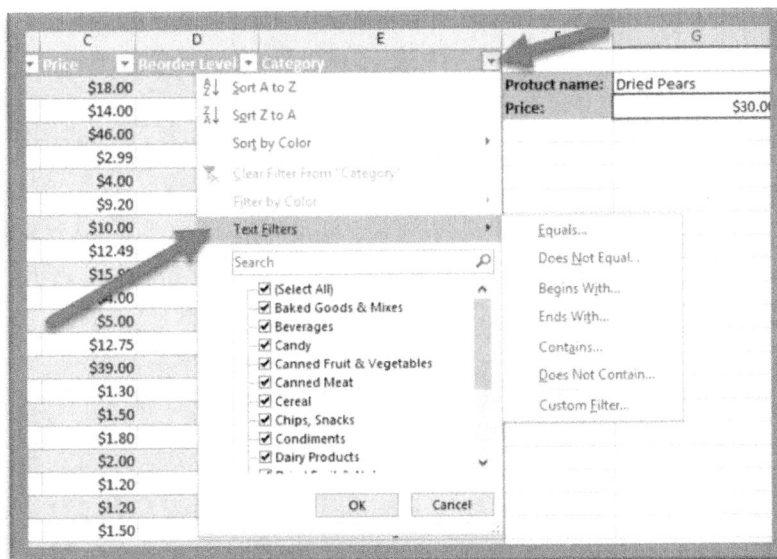

For example, if you wanted to display data with a price range between $2 and $10, you would: Click on the *Price* AutoFilter and then select *Number Filters > Between...* from the pop-up menu.

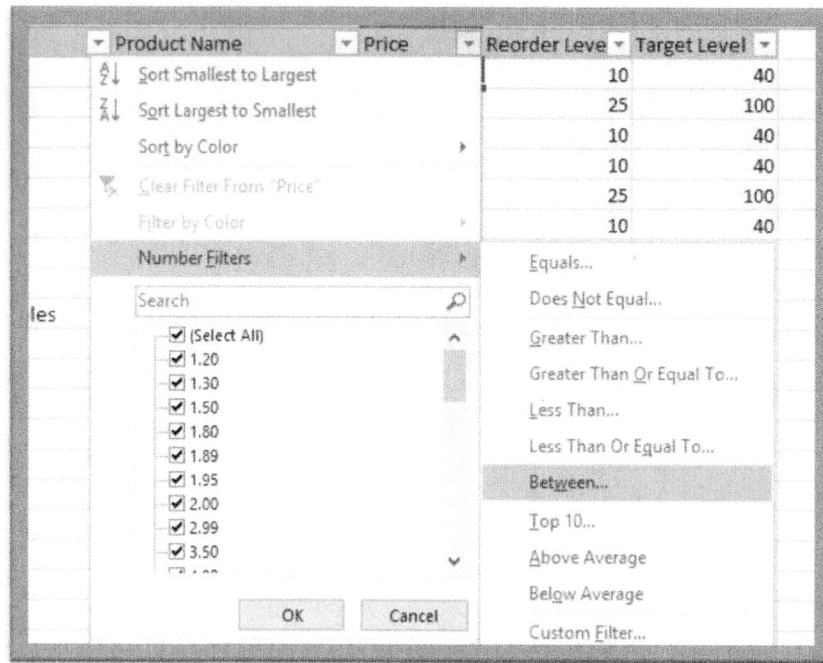

The *Custom AutoFilter* screen allows you to enter the criteria and specify the condition.

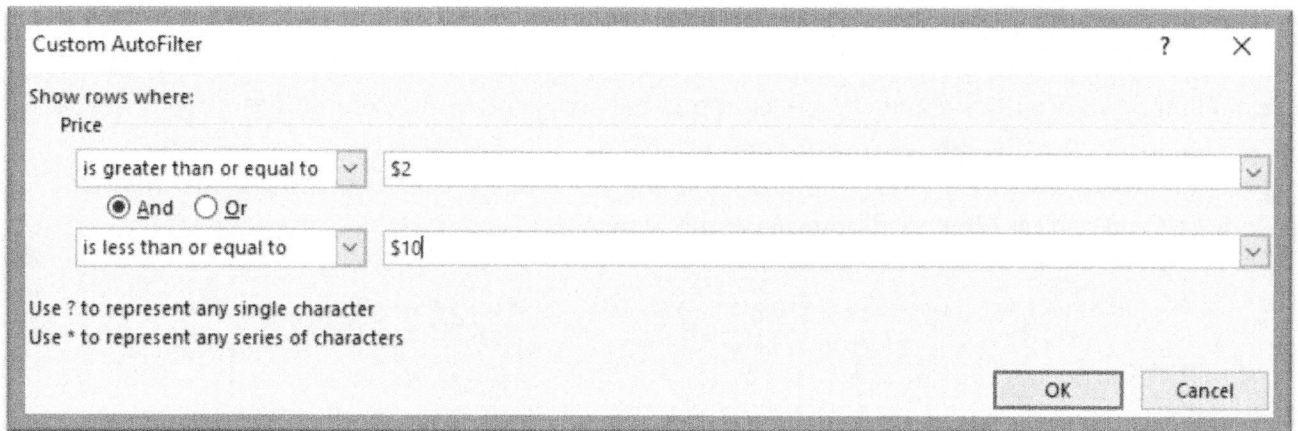

Enter the values you want to use for the filter. In our example, the values would be $2 and $10. Select the logical operator. In this case, we'll need *And* as both conditions must be true. Price >= $2 *And* <= $10. If only one of either condition needs to be true, then you would select *Or*.

Click OK when done.

The data will now be filtered to show only records with a price between $2 and $10.

How to Change the Sort Order of a Filtered List

Click the AutoFilter icon that appears on the column used for the filter to change the sort order of the filtered results.

Sort by largest to smallest or smallest to largest. Sort A to Z or Sort Z to A would be appropriate for a text column.

Removing a Filter. Select any cell in the range/table and click on *Clear* in the *Sort & Filter* group. The filter will be removed, and all data will be displayed.

Chapter #16: Matrix formulas

Index Match

The formula is given as: =INDEX (C3 ratio E9, MATCH (B13, C3 ratio C9,0), MATCH (B14, C3 ratio E3,0)). INDEX MATCH is a versatile Excel formula mix that can help you improve your financial research and modeling. INDEX is a table function that returns the value of a cell depending on the column and row number. MATCH returns the row or column direction of a cell.

Here's an example of combining the INDEX and MATCH formulas. We look up and return a person's height based on their name in this case. We should adjust both the name and the height in the calculation since they are both factors.

IF in Combination with AND/OR

The formula is given as: =IF (AND (C2 is greater or equal to C4, C2 is less or equal to C5), C6, C7). Anyone who has used a significant amount of time working with different financial models understands how difficult nested IF formulas can be. Combining the IF feature with the AND / OR function will make formulas simpler to audit and appreciate for other users. You will see how we combined the individual functions to construct a more advanced formula in the illustration below.

OFFSET in Combination with SUM or AVERAGE

The equation is as follows: =SUM (B4 ratio OFFSET (B4,0, E2 minus 1)). The OFFSET function isn't particularly complicated on its own, but when combined with other functions such as SUM or AVERAGE, we can create a complex formula. Consider the following example: You want to create a complex feature that can add a variable number of cells together. The average SUM formula can only perform static calculations, but you can shift the cell relation around by using OFFSET.

The way it works is as follows: We use the OFFSET function instead of the SUM function's ending comparison cell to make this formula work. This complicates the formula, and you can tell Excel how many sequential cells you want to sum up in the cell labeled E2. We now have advanced Excel algorithms. This much more complex formula is shown in the screenshot below.

CHOOSE

The formula is given as: =CHOOSE (Choice, option 1, 2, 3). The CHOOSE role is ideal for financial simulation scenario study. It encourages you to choose from a set of choices and can return the "choice" you've made. Assume you have three separate sales growth projections for next year: 5%, 12%, and 18%. If you tell Excel, you want option #2, and you will get a 12 % return using the CHOOSE formula.

XNPV and XIRR

= XNPV Formula (discount rate, cash flows, dates). If you work in investment management, market research, financial planning and analysis (FP&A), or another field of corporate finance that includes discounting cash flows, these calculations can be useful.

Simply put, XNPV and XIRR allow you to assign different dates to different discounted cash flows. The flaw in Excel's simple NPV and IRR formulas is that they assume that the time intervals between cash flows are equivalent. As an economist, you'll encounter situations where cash balances aren't evenly spaced on a regular basis, and this formula is how you solve them.

PMT and IPMT

Formula: =PMT (# of periods, interest rate, present value). You'll need to know these two formulas if you operate in real estate, commercial banking, FP&A, or other financial analyst jobs that deal with debt schedules.

The PMT theorem calculates the worth of making equivalent payments throughout a loan's existence. You should do that in combination with IPMT (which shows you how much interest you'll pay for the same kind of loan), then different principal and interest payments. Here's how to use the PMT feature to calculate the annual mortgage payment on a $1 million loan with a 5% interest rate over 30 years.

LEN and TRIM

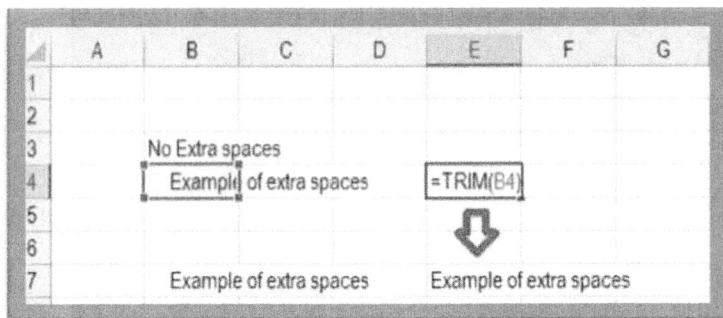

The following formulas are provided: =LEN & =TRIM (text) (text). The preceding formulas are less general, but they are more complex. They may be useful to financial professionals who need to manage and manipulate large amounts of data. Unfortunately, the data we receive is not always well-organized, and issues such as extra spaces at the start or end of cells will occur.

The LEN formula returns the number of characters in a given text string, which is useful when you need to count the number of characters in a text string.

You will see how the TRIM algorithm cleans up the Excel data in the illustration above.

CONCATENATE

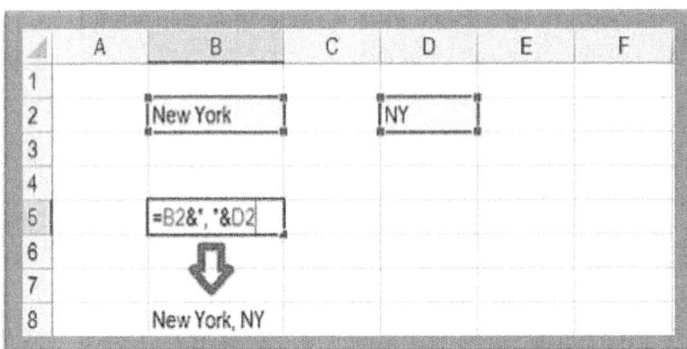

The formula is given as: =A1&"more text."

Concatenate isn't a function in and of itself, it's just a creative way of bringing data from multiple cells together and making worksheets more complex. For financial analysts doing financial simulations, this is a valuable instrument.

In the illustration below, the text "New York" plus "," is combined with "NY" to form "New York, NY." This enables you to make dynamic worksheet headers and labels. Instead of upgrading cell B8, you will also upgrade cells B2 and D2 independently. This is a great attribute to have while dealing with a vast data collection.

CELL, MID, LEFT, and RIGHT

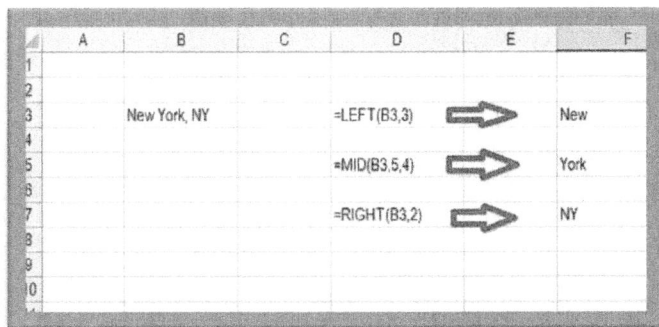

These complex Excel features can be combined to create specific, complex, and advanced formulas. The CELL feature may return a variety of data about the contents of a cell (such as its name, location, row, column, and more). The LEFT function returns the text from the cell's beginning (left to right), the MID function returns the text from any cell's beginning (left to right), and the RIGHT function returns the text from the cell's end (left to right) (right to the left). The three formulations are shown in the diagram above.

Chapter #17: Macro

A macro is a set of actions or an equation that can be used or repeated repeatedly. A macro assists in computerizing or repeating functions by recording or saving input patterns such as cursor strokes or keyboard clicks. After saving this information, it is used to create a macro that is accessible to all updates.

- Macros save time by recording, storing, and executing routine tasks as needed.
- The display tab contains macros. To record a macro, tap on it.
- Next, select the option to record macros.
- Pay close attention. Everything done with the spreadsheet is now saved.
- To stop recording, click the stop recording option that appears when one clicks the macros icon.

The primary motivation for developing a macro is the desire to automate routine tasks and avoid the continuous compilation of formulas suitable for the purpose. Furthermore, the fewer actions you perform in Excel, the less likely you are to make formula-setting mistakes.

Excel Macro is a tool that simply records and replays your Excel steps as often as you want. VBA Macros help you save time by automating repetitive tasks. Macros are pieces of programming code that run in an Excel environment, but you don't have to be a coder to create them. However, basic VBA knowledge is required to make advanced changes to the macro.

Record a Macro

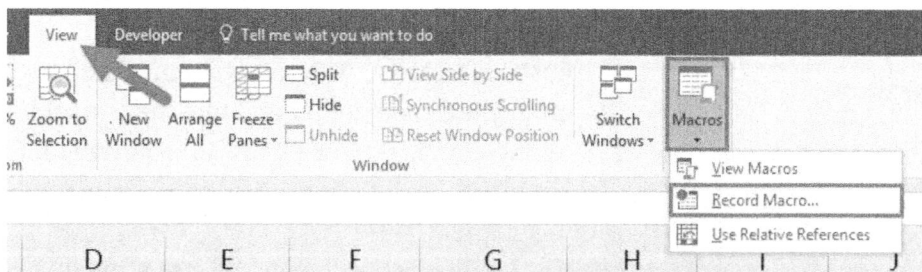

As always in Excel, what seems very difficult to explain in words, becomes a child's play with a demonstrative example. Two types of macros can be created: one direct and intuitive; the other indirect and reasoned. We will see the first method, which is also the most common among Excel users.

Consider that in our work, we frequently work on a variety of cells and transform them into their result. In order to achieve the same result in Excel, you must first select the range of cells, then make a copy, and finally the paste special values. If you have to do this frequently and on large worksheets, having a customized keyboard shortcut handy can be useful and convenient.

Macros are used to accomplish this.

First, we prepare the data by inserting some formula in the selected range, which could be anything, such as random (). We choose the entire range containing the data. Let's get started by following these steps:

- Go to the View tab, then select macros and record macros.
- Enter Convert Formula in the box where the macro's name is required.
- In the Shortcut text box, hold down the shift key and type Z. CRTL + Shift + Z will be our shortcut.
- We select this workbook, which we'll call macro1 because it's the first macro we create, and then click Store macro in.
- Leave the text in the macro description alone.
- After clicking OK and closing the dialog box, you will notice a button appear on the status bar indicating that all commands typed on the worksheet will be recorded from this point forward.
- All that remains is to make a copy; in the contextual menu that appears, select paste special, then paste special values; by clicking OK, we transform the formulas in the range of their values. We deactivate the interval by pressing the Esc key.

- Now, let's click on the recording icon in the status bar, which will change to indicate that the macro has been recorded.

Now all you have to do is try the macro: enter a formula in a new range, press the shortcut key you created CRTL + Shift + Z, and all formulas will be converted in the result.

Now we just have to save the workbook containing the macro, Excel will immediately suggest that the folder cannot be saved with the usual extension: .xlsz, but will have a name that makes the folder containing a macro recognizable with the extension: .xlsm.

Make a macro

If you have repetitive tasks in Microsoft Excel, you can record a macro to automate those tasks.

When you create a macro, you are recording your mouse clicks and keystrokes. After you've created a macro, you can edit it to make minor changes to how it works.

Assume you produce a monthly report for your accounting manager. Customers' names with past-due accounts should be highlighted in red and formatted boldly. You can quickly apply these formatting changes to the cells you select by writing and running a macro.

Before you begin recording a macro, enable the Developer tab in the ribbon, which is hidden by default and contains macros and VBA tools.

1. On the Developer tab, in the Code group, click Record Macro.
2. Before clicking OK to begin recording, optionally give the macro a name in the Macro name box, a shortcut key in the Shortcut key box, and a description in the Description box.
3. Perform the actions you want to automate, such as entering boilerplate text or populating a data column.
4. On the Developer tab, click Stop Recording.
5. Pay attention to the big picture.
6. You can learn a little about the Visual Basic programming language by editing a macro.
7. To edit a macro, go to the Developer tab's Macros group, select the macro's name, and then click Edit. The Visual Basic Editor is now available.
8. Examine how the recorded actions appear as code. Some of the code will almost certainly be obvious, while other parts may be more enigmatic.
9. Experiment with the code, then exit the Visual Basic Editor and rerun your macro. Watch to see what happens this time!

Enabling Macros When the Message Bar Appears

When you open a file with macros, a yellow message will appear asking you to enable the content. This is accomplished as follows:

1. Click on the Enable option in the Message Bar.
2. The yellow message that detects macros in the file will appear.
3. Enable the content to work with macros.

Enable Macros for the Current Session Only

While the file is open, the following instructions will enable macros:

1. Select the File tab.
2. In the Security Warning section, select Enable Content.
3. Go to the Advanced tab.
4. In the Microsoft Office Security Options dialog box, select Enable content for this session for each macro.
5. Press the OK button.

Changing Macro Settings in the Trust Center

When you change macro settings in the Trust Center, only the settings in the Office program you are using are changed.

1. Click on the File tab.
2. Click on Options.
3. Click on Trust Center and then on Trust Center Settings.
4. Click on Macro Settings.
5. Select the required options and OK.

Conclusion

Thank you for reading this book. Excel is an excellent tool for performing analyses and what-if scenarios. Microsoft Excel is a powerful spreadsheet and data analysis program with a wide range of capabilities. Data of various types can be organized, calculated, and saved for future use.

Excel is a spreadsheet-based program that has hundreds of features that can help you with your work. Excel is so popular these days that people who know it even on a basic level are paid more than people who don't know it at all.

The Excel database makes it easy to create, access, update, and share data with others. Excel skills are among those that have grown in importance in today's job market. They carry a lot of weight because Excel represents not just one type of skill, but a wide range of skills that employers are very interested in. As a result, people with strong Excel skills have a better chance of landing a job than those with little or no experience with the spreadsheet.

You can handle all tasks and projects on your Excel worksheet now that you've read this book. It is a valuable tool that you can use to properly manage your records regardless of whether you work in corporate organizations, financial institutions, agricultural sectors, or a private business enterprise.

Excel is used to complete daily tasks by accountants, investment managers, consultants, and individuals in all aspects of financial careers. Microsoft Excel will remain the most widely used framework for analyzing data, creating maps and presentations, and integrating with powerful software for digital dashboards and business intelligence workflows.

Excel skills will make your life much easier in the workplace. For example, you can easily gather data if necessary, analyze it if necessary, and draw some conclusions from the data.

Mastering these fundamental Excel skills is what you need to do in order to make your job easier and impress people around you at work. Remember, no matter how familiar you are with this useful tool, there is always something new to learn about its operation. Whatever you do, continue to hone your Excel skills. Excel is designed specifically for you - you don't want to waste time doing things that Excel 2023 can do in a fraction of the time. You should strongly advise your family, friends, and coworkers to use it.

Overall, Microsoft Excel simplifies the manipulation, interpretation, and analysis of data, allowing you to make better decisions while saving time and money. Microsoft Excel provides the tools you need to get the job done, whether you're working on a commercial project or managing your personal finances. Excel 2023 will allow you to take your business, profession, and other endeavors to new heights.

Hopefully, after reading this book, you will realize that this guide is excellent for starting your journey with Excel and creating your workbooks. This guide contains various methods and techniques for dealing with Excel and beginning to guide others who are beginners and new to MS Excel.

MS Excel is a simple software program, and understanding the fundamentals will help both beginners and experts advance in their careers. Beginners may be more concerned with simple functionality such as columns, rows, and tables, and they may be less knowledgeable about the program's enhanced features. To use the program in your day-to-day workplace tasks, you must first understand the application and its benefits.

Mastering these fundamental Excel skills will make your life easier — and may even impress those in your workplace. However, keep in mind that no matter how familiar you are with this useful tool, there is always something new to learn. Whatever you do, continue to improve your Excel skills — they will not only help you keep track of your own earnings, but they may also lead to a better potential job opportunity.

This book is your ultimate Excel guide, and it will help you learn and work with Excel without any complications. This book is an easy-to-follow guide for everyday use, and it may help you learn this program quickly whether you are a student, working, or retired.

This book has covered all of the advanced Excel features and strategies that one would find useful in speeding up existing tasks or optimizing data as well as providing new tools and insight into certain features that provide them with a lens through that they can manipulate and analyze their data and control information from an entirely new perspective.

Excel is often unavoidable in marketing, but with the above tips, it doesn't have to be as intimidating. As the saying goes, practice makes perfect. These formulas, shortcuts, and methods will become second nature as you use them more frequently. If you enjoyed this book, I would appreciate it if you could leave a review.

Thank you and Good Luck!

Bibliography

- https://techcommunity.microsoft.com/t5/excel-blog/what-s-new-in-excel-june-2022/ba-p/3517410
- https://www.udemy.com/course/learn-microsoft-excel-/
- https://clickup.com/blog/google-sheets-vs-excel/
- https://lifehacker.com/seven-useful-microsoft-excel-features-you-may-not-be-us-1753221566
- Schonlau, M., & Peters, E. (2012). Comprehension of Graphs and Tables Depend on the Task: Empirical Evidence from Two Web-Based Studies. Statistics, Politics, and Policy, 3(2). https://doi.org/10.1515/2151-7509.1054
- https://www.microsoft.com/en-us/microsoft-365/excel

Printed in Great Britain
by Amazon